Make It Glow

LED PROJECTS FOR THE WHOLE FAMILY

In the daytime, the world is a colorful place.
At night, darkness may steal those colors away.
But with the activities in this book—along with
a little creativity and inspiration—
you can light up your world, and make it glow.

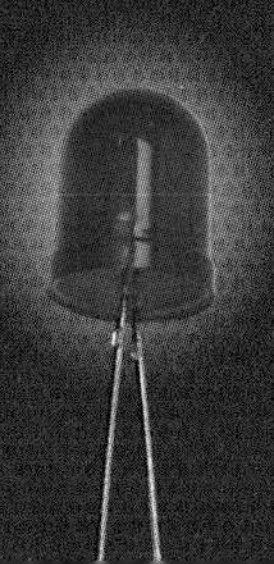

Make:

Make It Glow

LED Projects for the Whole Family

Emily Coker and Kelli Townley

Printed in Canada.

Published by Maker Media, Inc., 1160 Battery Street East, Suite 125,
San Francisco, California 94111.

Maker Media books may be purchased for educational, business, or sales promotional use. Online editions are also available for most titles (safaribooksonline.com). For more information, contact our corporate/institutional sales department: 800-998-9938 or corporate@oreilly.com.

Publisher: Roger Stewart

Producer: Leslie Jonath, Connected Dots Media LLC

Editor: Ruth Tepper Brown

Photographer: Rory Earnshaw

Designer: Kevin Plottner

Copy editor: Jeff Campbell

September 2016: First Edition

Revision History for the First Edition

2016-09-01 First Release

See oreilly.com/catalog/errata.csp?isbn=9781680451054 for release details.

978-1-68045-105-4 (2016-09-16)

TABLE OF CONTENTS

CHAPTER 3: Soft Circuitry & Wearables .... Page 57

Power up your wardrobe with activities that pair sewable LEDs with conductive thread. Make everything from trendy fashions to heartwarming stuffies, and try out new tools and techniques to expand your DIY horizons.

CHAPTER 4: Twist, Solder, Tape & Hack .... Page 103

Mix it up! Once you get a feel for all the ways you can use LEDs, batteries, and circuits to make things glow, you can begin experimenting with different ways to put them together.

INTRODUCTION

Many of the most amazing devices around us today—from microwave ovens that can sense when your potato is done to the videogame systems that can respond to your spoken commands—get their smarts from the electronics inside.

Electronics (see the word "electron" in there?) is essentially the science of manipulating electrical energy. And that's where this book begins. These activities have been crafted to give everyone—from the tech-savvy to the tech-wary—a way to start exploring the skills and materials that are at the heart of the world of electronics.

The projects in this book introduce you to some simple, fun ways to brighten your life. With these activities, you can light up the night with LED balloons, make a glow-in-the-dark fashion statement (for you and your pet!), and create lightning in your living room. Try making soda-bottle ghosts with glowing eyes, crafting light-up holiday cards, and bringing a well-loved stuffie to life.

It's Not Just Cool... It's Science

Everything in the world—including you—is made from zillions of tiny particles called atoms. No matter how nearsighted you are, you can't actually see them, so don't bother trying. But by using special microscopes, scientists can see that they're there.

Moreover, each atom is made up of even tinier bits and pieces called protons, neutrons, and electrons. Put those pieces together in different ways, and you get all the different elements in the universe—from the carbon in your pencil lead to the calcium in your bones. They're all made out of atoms.

You can think of an atom as a little clump of protons and neutrons surrounded by a swarm of electrons madly zipping around it. The way those electrons behave is what we'll be exploring in this book. By manipulating them with magnets and batteries, and moving them through wires, threads, and tapes, we're going to make the world glow.

This is not a technical treatise. You don't have to know anything about electricity to create these projects. You don't need to be an electronics wiz or a computer geek. In fact, you don't even need a computer. You just need to be willing to play around with stuff you may never have explored before.

Be sure to gather the tools and materials for each project before you begin. You should be able to find everything you need online or at stores that sell crafts and party supplies, hardware, and electronics. We've tried to help out by including photos of some of the most important materials in each activity so you can identify components, even if they're unfamiliar.

You can do the activities in any order, but we recommend beginning with chapter 1, which introduces some basic ideas and techniques. Some projects are more challenging than others, so we've also included a handy skill-level guide, which you can find on page 133.

That said, these projects do require a few basic skills and techniques:

Chapters 1 and 2 introduce easy activities for lighting things up by using LEDs and batteries, and you'll make simple switches. No special skills are needed, though parental guidance may be a good idea for children who have not had practice with hot-glue guns or craft knives.

Chapters 3 and 4 introduce new tools and more-advanced techniques. Household tools such as wire strippers, needle-nose pliers, and screwdrivers are occasionally needed. Several activities require sewing and soldering. If you've never used a needle and thread or handled a soldering iron, be sure to have an experienced Maker with you who can show you the ropes—or do your homework to learn how to use these tools yourself, so you don't get hurt!

For your convenience, skill levels for each activity are identified so you know what you're getting into. For activities sorted by skill level, see the list on page 133.

SKILL LEVEL 1

FOR THE FUN OF IT

No special skills or experience are required for these projects, though hot-glue guns, craft knives, and other common tools and materials may be needed.

SKILL LEVEL 2

FOR YOUR INNER SCIENTIST

These intermediate-level projects introduce new materials, tools, and techniques—from paper folding (origami) to simple sewing skills—and they may require participants to follow more precise instructions. Adult supervision suggested.

SKILL LEVEL 3

FOR THE ADVENTUROUS MAKER

These more challenging projects may take longer or require more complex skills, including sewing and soldering. Some introduce new components, such as resistors. Adult supervision highly recommended.

A PRACTICAL PRIMER: CIRCUITS, BATTERIES, LEDS, AND MORE

The materials needed to explore the world of electronics may not be familiar to work with, but you've seen them, and used them, all your life. Whenever you use your cellphone, printer, or computer, flip on a fan or light switch, or play a videogame, there are electronics at work, making things go.

Most of us never get to see how the circuits, batteries, bulbs, motors, LEDs, and wiring inside our electronic devices actually work, but those are exactly the kinds of things we're going to play with here. This primer will give you a simple, nontechnical foundation that will help you start making discoveries on your own.

Tips for Creating the Circuits Used in This Book

- Your job in these projects is to connect batteries to LEDs via one-way, circular pathways, called circuits. The batteries provide electrical current; the LEDs use it.
- The metal wires, threads, and tapes you use to connect your batteries and LEDs are known as "conductive" materials: they "conduct" the flow of electrons through circuits. Once electrons are flowing through a circuit, they make LEDs light up. (Like flowing water, flowing electrons are also called a current!)
- Always make sure you've created a clear and complete path for current to flow through your LEDs. If the electrons in your circuit flow directly from one side of your battery to the other (without going through an LED or other electrical component in between), they cause a problem called a "short circuit"—or sometimes just "a short" ... for short. Without something to power (such as an LED), too much current will flow, causing damage to your circuit. Short circuits can happen if the current isn't doing anything but circulating back to the battery, or if parts of your circuit cross over one another. Check your battery—if it's heating up, it's a sure sign of impending problems.

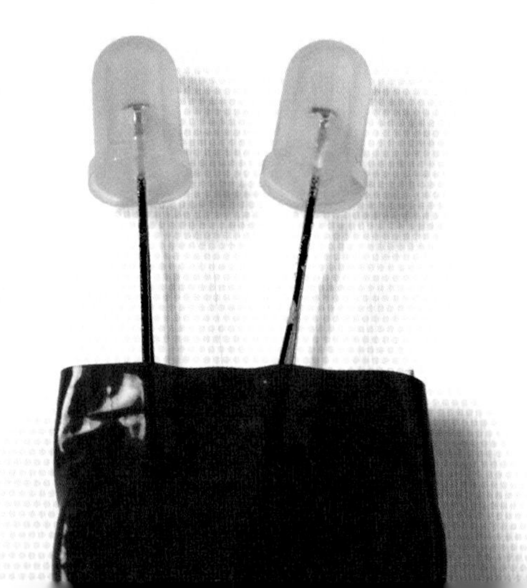

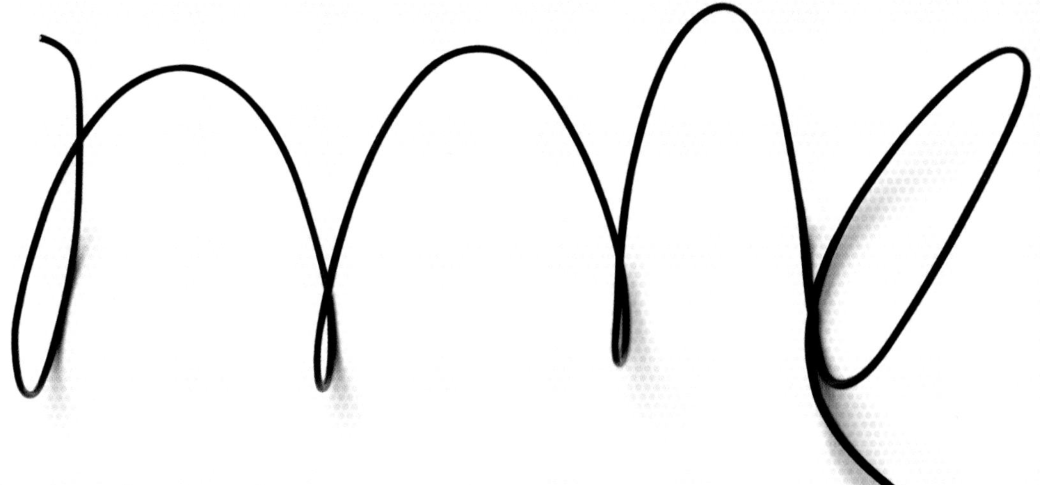

Tips for Using Batteries & Resistors for the Projects in This Book

- All but one of the projects in this book use small, button-sized, 3-volt, lithium coin-cell batteries, identified commonly as CR2032. (The Light-up Tote Bag uses a 9-volt battery instead.) You can buy packs of coin-cell batteries just about anywhere; get them a few at a time or in bulk. But don't jumble them together. Regular batteries have their positive and negative terminals on the ends, so it's easier to keep them apart. A coin-cell battery's terminals are exposed. If they touch, they can short each other out. Either keep coin-cell batteries in the package or tape them individually to pieces of cardboard so they stay fresh.
- Batteries generate voltage. LEDs use it up. In any circuit, the two must balance. To make sure each project works as intended, we've figured out which batteries work best with which LEDs, so be sure to use the ones recommended! If you double up on batteries to make an LED shine brighter, you're just likely to burn it out instead.
- You need to "feed" your LEDs just the right amount of electrical current for them to be at their best. Not enough, and they won't glow. Too much, and they'll burn out. In a few projects, we've added resistors to control the amount of current getting to the LEDs. They'll help the batteries last longer, protecting your circuit from any electrical overloads and letting each LED shine its brightest. The 3-volt coin-cell batteries that power most of these projects create low levels of electricity, so can be safely used without resistors.
- Coin-cell batteries are great for projects, but the chemicals inside are highly corrosive, and their small size makes them attractive to toddlers and young children. Be sure to store and dispose of your batteries responsibly. Keep them far away from small children, and don't just throw them in the trash when you're done with them! Many hardware and crafts stores will recycle them for you, or check with your local refuse company to find out if they will take spent batteries.

Tips for Using LEDs for the Projects in This Book

- All of the projects in this book glow by using LEDs, or "light-emitting diodes." LEDs come in lots of sizes and colors. Some LEDs blink, some pulse, some cycle through a rainbow of colors. Some LEDs are blindingly bright, and others glow softly, with diffused light. There are flat "surface mount" LEDs, bulky "gumdrop" LEDs, adhesive "circuit sticker" LEDs, sew-on "sequin" LEDs, and even long strips of "ribbon" LEDs. Specific sizes (3mm, 5mm, 10mm, for instance) refer to the diameter of the LED. The variety is endless, and different kinds are available online and at craft stores.
- We've recommended specific LEDs for some of these projects, but you can pick your favorites for most projects. As you work, though, note that different LEDs use different amounts of current, so some colors or sizes may end up being brighter than others. Experiment to find the ones that work best for you.

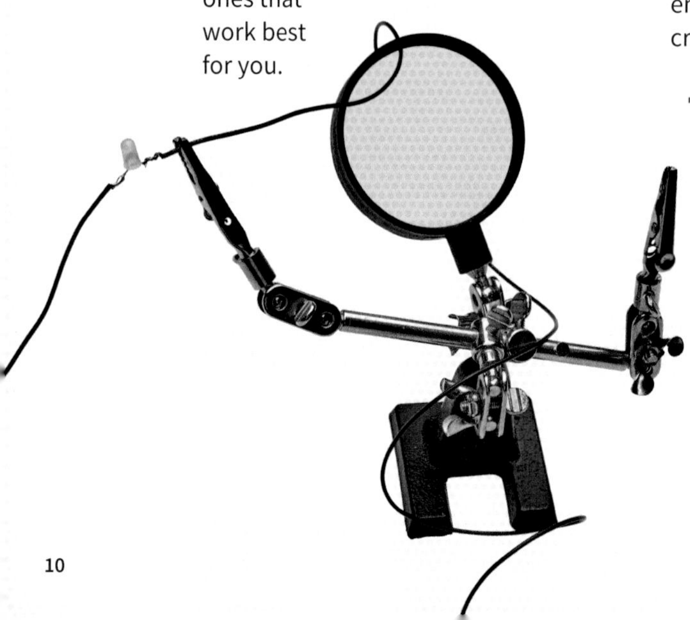

Tips for Using Conductive Materials for the Projects in This Book

- Where wire is used in these projects, look for stranded wire in the 18–22 gauge range. And while you might think that electrical current can only flow through wires, that's not the case! These days, there are lots of cleverly made conductive materials you can use in circuits. In addition to wire, we've used copper tape and conductive thread, but you might want to experiment with conductive paints, glues, and inks, too.
- Note that there are lots of different types of copper tape on the market. Some are stickier than others; some are more conductive than others. They also come in different widths, but you can cut copper tape lengthwise if it's too fat to fit neatly in your projects. A few tear too easily to be useful for electronics projects, and some have coatings that need to be scraped off at their terminal ends. Watch for products made for crafts and Maker usage.
- Conductive thread, in general, can be knotty stuff to use—after all, it's made of metal. But the better the quality, the easier it is to work with. Other than that, you can just sew with it like you would any thread. You don't need to double it, like you would in some sewing tasks; single threads will work best.

ABOUT BEING A MAKER

We're all Makers, of course—creators of personal, imaginative, unexpected stuff that may come out of a classroom, kitchen, office, playroom, sewing room, or just about anywhere. But the community of Makers who experiment with electronics take their creative juices one step further, playing with the very nature of, well, nature itself.

The team that worked on this book brainstormed and experimented at every step and ended up taking some unexpected paths. Now it's your turn. If you think something would work better upside down and inside out, give it a try! If you want to see what happens when you use old Christmas lights instead of flashing LEDs, use them! If you think sticky-back Velcro will work better than glue, Velcro away!

Let your curiosity drive your creativity, and you'll find that one discovery will naturally lead to another. And if you find yourself suddenly inventing your own projects, then you've joined the ranks of Makers around the world, and our job here is done.

AN IMPORTANT NOTE ABOUT SAFETY

Children working on the activities in this book will often need adult supervision. Depending on the project, participants may be asked to use craft knives, hot-glue guns, needles, soldering irons, and other tools and materials that will require oversight. Please keep safety in mind as you experiment and explore.

Maker Media connects Makers with one another, creating ideas and products that let curious inventors push the boundaries of their own knowledge and skills.

Learn more about us at makermedia.com.

CHAPTER 1
Glowies & Throwies

A tiny LED, a little round battery, and a strip of tape are all it takes to add dazzle in the most unexpected places. These little glowing lights, called Glowies, can light up balloons, reveal secret messages, or create the sinister eyes of a Halloween ghost.

Begin with Start by Making Glowies, page 14, and everything in chapter 1 will make perfect sense. Choose LEDs that blink, pulse, cycle through a rainbow of colors—whatever you want!

Once you know how to make a Glowie, you can glow up pretty much anything. So keep your batteries, LEDs, and tape handy, and you'll be an expert in no time.

Start by Making Glowies

SKILL LEVEL 1

Tape an LED to a battery and you've made a "Glowie." This simple circuit is the secret ingredient for all the activities in this chapter. In fact, it's the basis for lots of different ways to make your world glow.

Get your tools & materials...

TOOLS

- Scissors

MATERIALS

- One LED (any type)
- One 3V coin-cell battery (CR2032)
- Tape (any type)

...and MAKE it!

1. Sandwich a battery between the legs of an LED. Make sure the LED's longer, positive (+) leg is on the smooth, positive (+) side of the battery, and its shorter, negative (–) leg is on the bumpy, negative (–) side of the battery. The LED should light right up. If it doesn't, flip the battery and try again.

2. Use scissors to cut a piece of tape, and tape the LED legs and battery tightly together.

Make Throwies

Created in 2005 by Graffiti Research Labs, Throwies are Glowies with magnets attached. They were designed to be used in public art displays, thrown onto metal surfaces hundreds at a time.

To make a Throwie, add a small (but powerful) neodymium ring or button magnet to one side of a Glowie (it doesn't matter which side). Tape the whole assembly tightly together, and you'll be ready to make some magnetic magic.

CR2032

LED + battery + magnet = Throwie!

People have used Throwies to decorate buildings, bridges, cars—anywhere they'll stick! You should be able to find lots of places to stick Throwies around your house, too, from decorating the spinning metal blades of a fan to spelling out messages on a refrigerator door.

Make Double or Triple Throwies and Glowies

Depending on your materials, you can tape up to three LEDs onto a single 3-volt battery and get several hours of light from the assembly. A single Glowie or Throwie can last up to three days. A double or triple Glowie or Throwie won't last as long, but it can make things a whole lot more colorful! The more LEDs you add, though, the dimmer they'll be, since they share one battery. Experiment to see what works best for you.

Once you know how to make Glowies and Throwies, you can use them to light up just about anything! Wear them on your head. Add sparkle to homemade decorations, blinking lights to cardboard robots, or flickers of faux fire to a roaring toy dragon. The possibilities are endless.

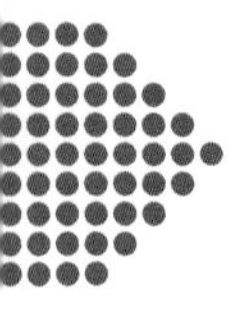

Floaties: Light Up the Night!

SKILL LEVEL 1

Floaties are simple and dramatic—all you need are some balloons and a lot of hot air.

Get your tools & materials...

TOOLS

» Scissors

MATERIALS

» One or more Glowies, enough for each balloon
» Balloons
» String
» Optional: helium tank (available in crafts and party stores)

...and MAKE it!

1. Use scissors to trim the tape on each Glowie so there are no sharp edges. Then carefully wiggle at least one Glowie into the neck of each balloon. (You may need to stretch the balloon neck with your fingers a bit to help it fit.) If you want, put more than one Glowie in each balloon or use double Glowies.

2. Blow up each balloon. It's easiest to use your own hot air, but if you want your Floaties to fly high on their own, use a tank of helium.

3. When each balloon is full, tie off the neck. Then cut lengths of string and tie them to each balloon. String lets you hang regular balloons or tie down helium balloons, so your Floaties don't float away.

Choose different kinds of balloons and make Glowies with a variety of LEDs to get everything from flashing pinpoints of light to spheres that glow softly in a darkened room.

Glow Ghosts

SKILL LEVEL 1

Wondering what just went bump in the night? The glowing eyes of soda-bottle ghosts are a dead giveaway.

Get your tools & materials...

TOOLS

- Scissors or craft knife
- Optional: hot-glue gun and glue sticks

MATERIALS

- Clean, empty, transparent plastic bottle (any size), with the cap, one for each ghost
- One double Glowie for each bottle, with the LEDs set next to each other (for the eyes)
- Transparent tape
- Length of clear plastic line or wire
- Square of lightweight white cloth, one for each ghost, big enough to drape over and completely cover the bottle

...and MAKE it!

MAKE THE GHOST

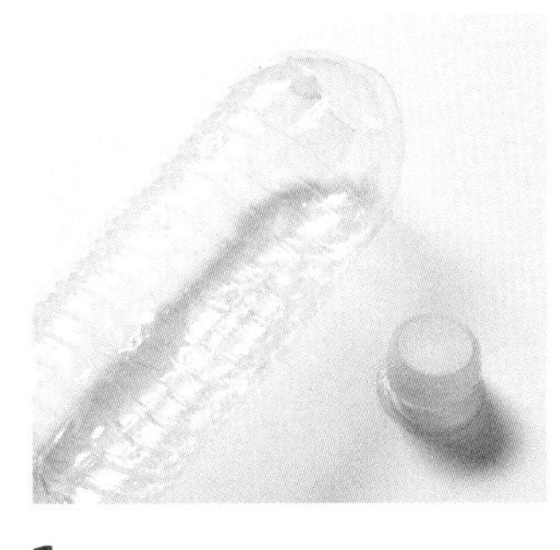

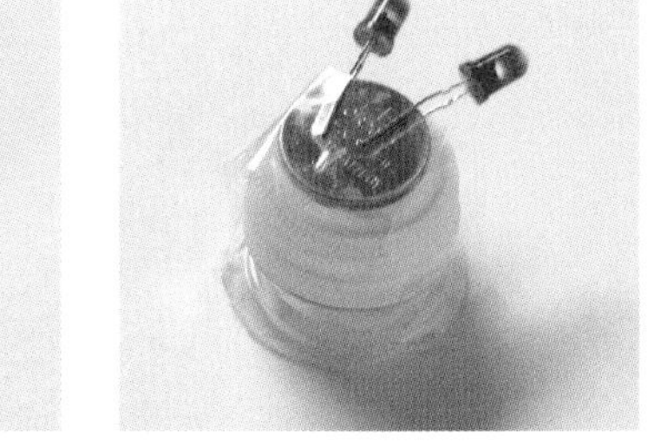

1. With your scissors or craft knife, carefully cut off the top of the bottle (with the bottle cap on!), just below the neck.

2. To make the eyes, tape or hot-glue the double Glowie on top of the bottle cap.

Now hang your Glow Ghost in a dark, scary place.

Ghoulish guardians of the night, Glow Ghosts were inspired by the Ghosties created by Evil Mad Scientist Laboratories.

3. Flip the bottle-cap assembly upside down and insert it into the cut opening in the bottle. The cap—with glowing LED eyes attached—will be inside. Tape the cap in place, making sure it doesn't fall into the bottle.

HANG IT UP

4. Cut a length of clear plastic line or wire. Poke it down through the center of the cloth, and then up again an inch or so away. Knot the ends, making a single loop attached to the fabric. Then tape or hot-glue the knot to the top of the bottle and pull down the cloth so it covers the bottle.

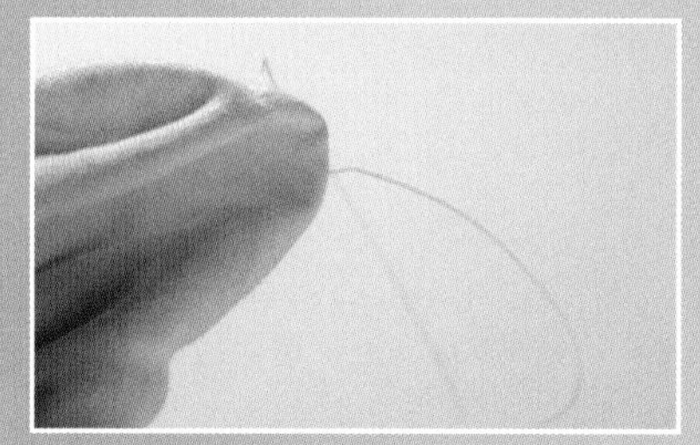

Eyes in the Bushes

SKILL LEVEL 1

Looking for an easy way to add a little excitement to your night? Wiggle these eerie eyes into the bushes in your yard, or hide them around the house to creep out family and friends. (Plus, it's a great way to use old toilet-paper tubes.) Be aware: If you put out a bunch on Halloween, expect to lose a few to admirers. That's what happened to us.

Get your tools & materials...

TOOLS

- » Scissors or craft knife

MATERIALS

- » Empty toilet-paper tube, one for each set of eyes
- » Roll of blue painter's tape
- » One double Glowie for each tube, with LEDs sticking out on opposite sides

...and MAKE it!

FOLD AND CUT

1. Use scissors or a craft knife to cut along the length of the toilet tube. Flatten the cut tube with your hand, so the tube is folded in half lengthwise.

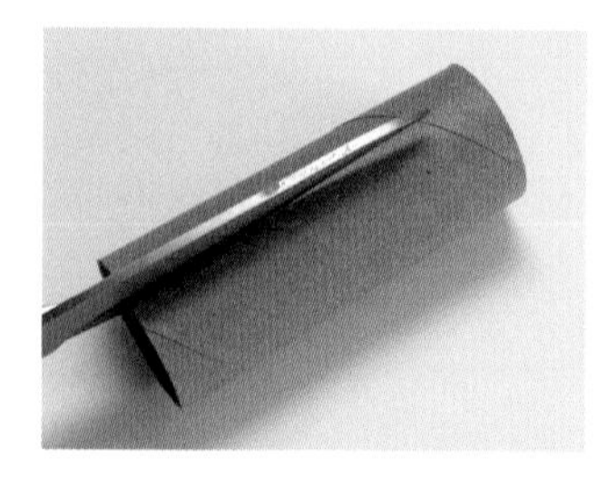

2. Bend or fold back the cut edges so they meet the middle fold. The tube now has four folded sections. In one of the middle sections, use scissors or a craft knife to cut a pair of creepy eyeholes at scary angles or in different shapes. If you're using scissors, another way to cut the eyeholes is to fold one middle section again and cut along this edge, as shown.

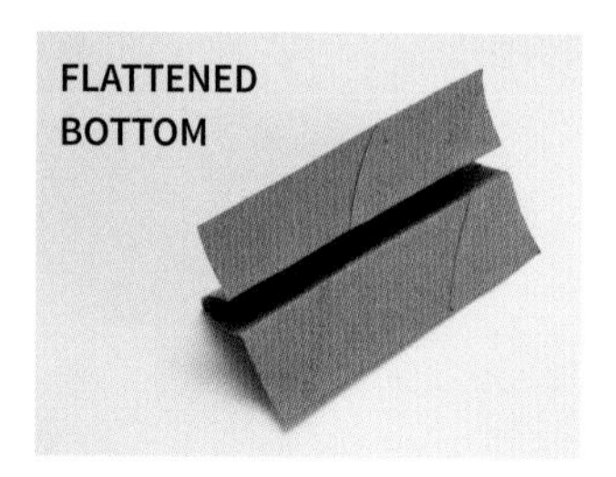

FLATTENED BOTTOM

UNCUT TOP

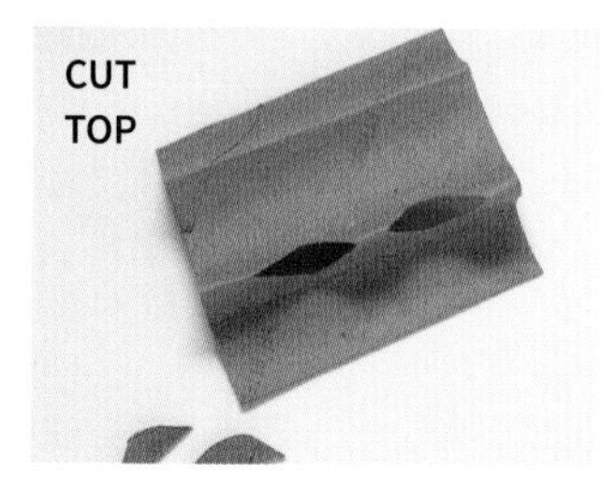

CUT TOP

CREEP IT UP

3. Using blue painter's tape, attach the double Glowie to the roll so that the LEDs appear in the eyeholes, as shown.

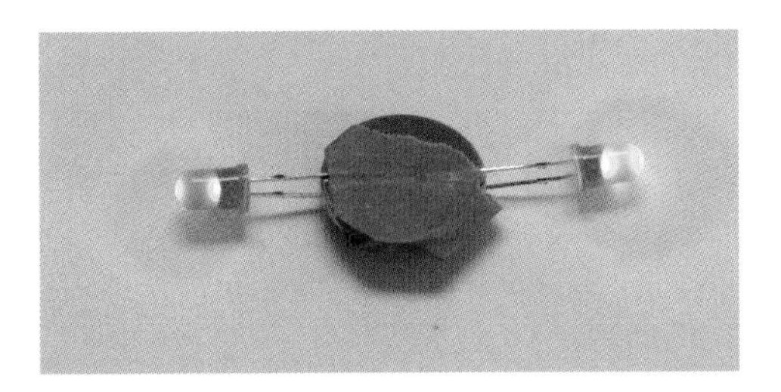

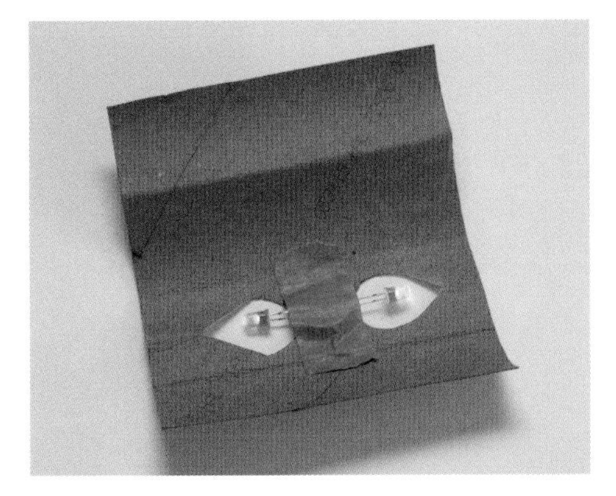

4. Fold the middle sections of the tube together, and hold them in place by covering the outside—including the eyeholes—with a single layer of blue painter's tape. Don't tape the tube's outer folds.

If your LEDs aren't shining brightly enough, just bend them a bit to shine directly through the eyeholes before taping.

5. Bend the outer folds to create a flat bottom, so your assembly can stand on its own. Then set the eyes in place indoors or wiggle them into the bushes to creep out unsuspecting guests.

So ghoul!

Flickerbugs

SKILL LEVEL 1

Flickerbugs are social creatures. Make a swarm.

Get your tools & materials...

TOOLS

- Ruler
- Wire cutters
- Needle-nose pliers
- Optional: scissors (to cut the tape)

MATERIALS

- Bendable craft wire or thin chenille stems ("pipe cleaners")
- A long, flat pencil eraser or a piece of floral foam, one for each Flickerbug
- One double Glowie for each bug, with two LEDs set next to one another (for eyes)
- Tape (any type)
- Fat, fuzzy chenille stems
- Optional: googly eyes

...and MAKE it!

MAKE THE BODY

1. Using a ruler and wire cutters, measure and snip six 6" pieces and two 2" pieces of craft wire or thin chenille stems for each Flickerbug. The 6" pieces are for legs, and the 2" pieces are antennas.

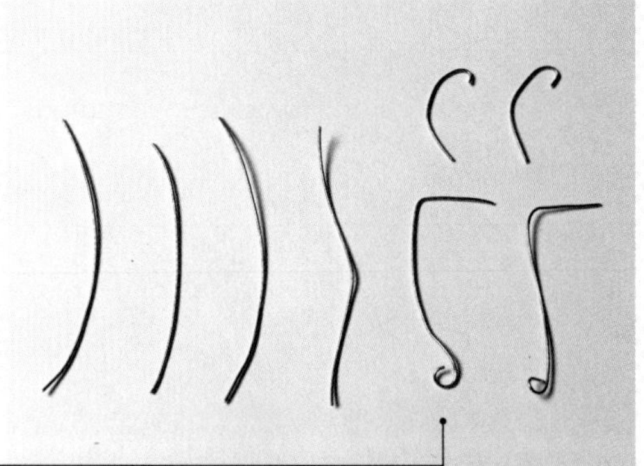

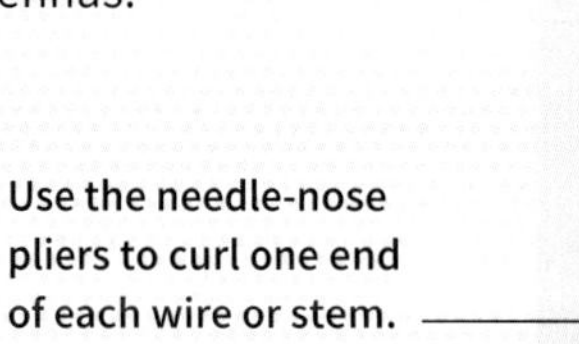

Use the needle-nose pliers to curl one end of each wire or stem.

2. To attach the legs, poke the straight ends of the long wires into the sides of the eraser (three per side), and bend each wire down so that the curly ends become feet. To attach the antennas, poke the straight ends of the short wires on top of one end of the eraser.

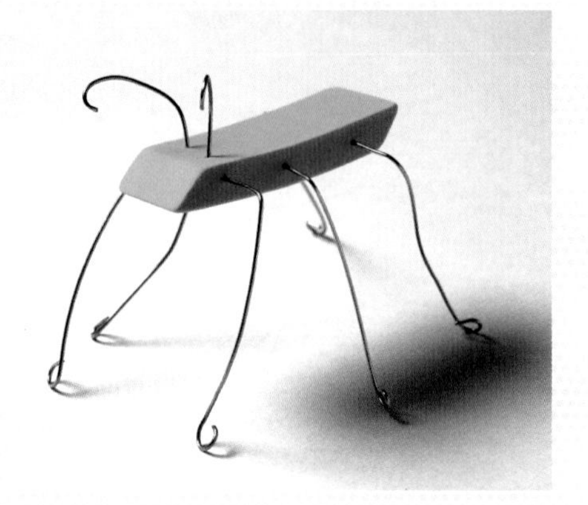

Flashing Flickerbugs are just one species of Blinky Bugs, creatures originally created by maker Ken Murphy.

BRING YOUR FLICKERBUG TO LIFE

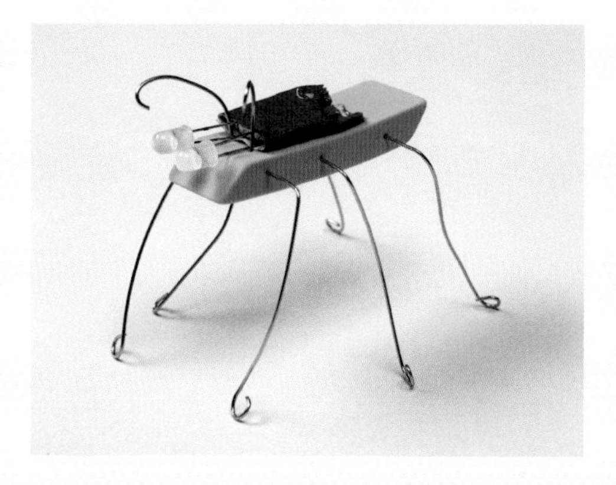

3. Place the Glowie eyes on top of the eraser, between the antennas and extending past the Flickerbug's body. Tape the Glowie lengthwise to the eraser to hold it in place. (The wings will help hide the assembly.)

4. To make wings: Using wire cutters, cut a fat chenille stem to the desired length and bend it in half. Fasten the stem to the Flickerbug's back with a piece of tape wiggled into the fuzz, and you're all done!

Want to make more? Try different-shaped bodies, or make spidery bugs with eight legs instead of six. Add black-paper batwings, toothpick stingers, googly eyes—just about anything you can think of. They're your bugs.

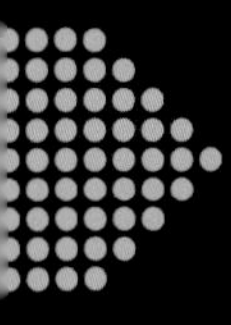

Awesome Ornaments

SKILL LEVEL 1

The whole world lights up during the holidays, and you can join the fun with these easy-to-make ornaments. They're beautiful hanging in windows, decorating trees and mantels, or tied to gifts. But that's not all! Fillable acrylic ornaments come in all different shapes and sizes, from bells, hearts, and teardrops to stars, globes, and more. Tie them around the necks of jars, vases, and bottles to sparkle up gifts of food, flowers, and beverages, or just make them for the fun of it, any time of the year.

Get your tools & materials...

TOOLS

- Scissors

MATERIALS

- Fillable acrylic ornaments (available online or at craft stores)
- Polyester batting
- A single or double Glowie (pages 14–15), one for each ornament
- Decorative string or ribbon

...and MAKE it!

1. Open the ornament, fill each side with batting, and wiggle a Glowie into the middle. Then snap the ornament back together.

2. Using the scissors, cut a length of string or ribbon, tie it through the loop on top of the ornament, and you're ready to glow!

Throwie Dartboard Game

SKILL LEVEL 2

A friendly game of magnetic darts can put the spark in a dark, dreary day or light up a party at night. Make your own board, your own darts—even your own rules.

Get your tools & materials...

TOOLS

- Black marker
- Scissors
- Wire cutters

MATERIALS

- Round 9" or 10" pie tin (make sure it's magnetic)
- Masking tape or painter's tape
- Old newspapers, a sheet of plastic, or other paint-proof covering
- Spray paint
- Six 3mm diffused LEDs—three of one color, and three of another color
- Copper tape
- Six 3V coin-cell batteries (CR2032)
- Electrical tape
- Six small neodymium magnets
- Duct tape

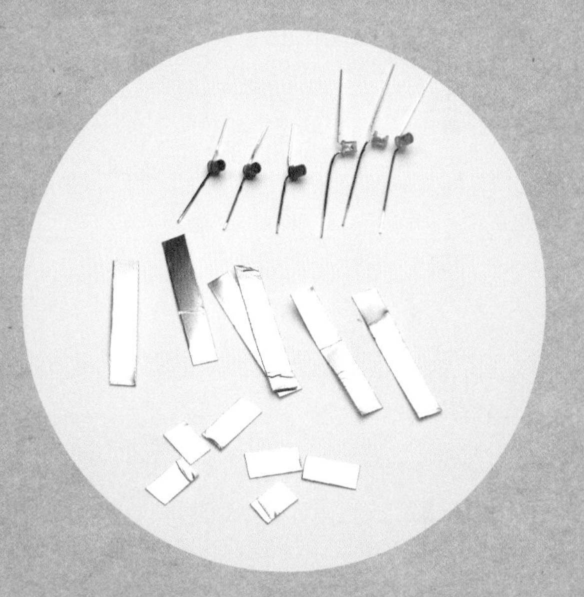

...and MAKE it!

MAKE THE BOARD

1. On the inside of the pie tin, make a cross with the masking tape, so you create four pie-shaped wedges. Press the edges of the tape to make sure they're stuck tight.

2. Lay down old newspapers, a sheet of plastic, or whatever you need to spray paint safely. (Get an adult to help with this!) Then spray paint the inside of the pie tin; several light coats will work best. Set aside to dry.

3. When the paint is totally dry, carefully peel up the tape. You now have a "dartboard" pattern like the one shown here.

MAKE THE DARTS

4. With the black marker, darken the long positive (+) leg of the six LEDs, and bend both LED legs to the sides.

5. Using scissors, cut six pieces of copper tape, each 1.5" long. Then cut six shorter pieces, each 0.5" long.

6. Put one of the short pieces of copper tape over the positive (+) leg of an LED, securing it to the positive (+) side of a battery, as shown. Then wiggle a small piece of electrical tape under the negative (–) leg of the LED, and wrap it around the edge of the battery (to keep the leg from touching the battery, so the circuit doesn't short out).

7. Take one of the long pieces of copper tape and fold it lengthwise over the end of the negative (–) leg of the LED.

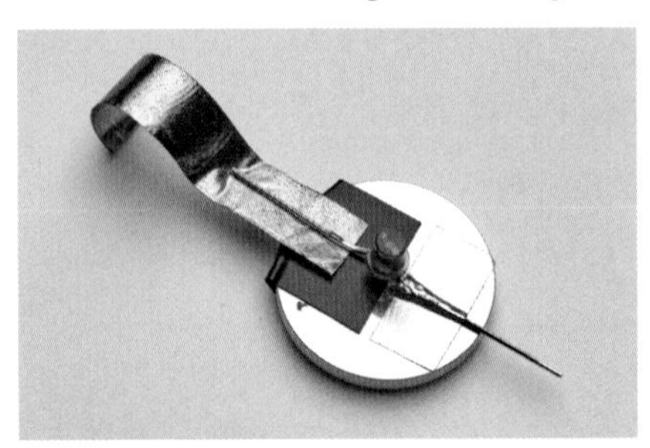

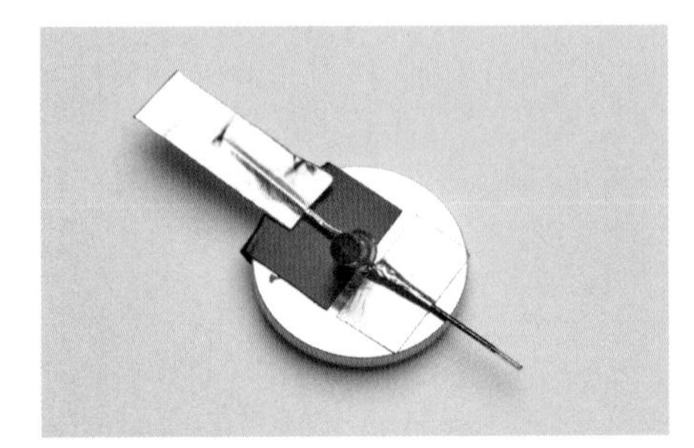

8. Carefully bend the negative (–) leg of the LED down and around to the back, negative (–) side of the battery. Secure with electrical tape. Turn the assembly over, and with the wire cutters, snip off the end of the positive (+) leg of the LED so it doesn't stick out beyond the edge of the battery. (Be sure you don't wrap the copper tape around the live edge of the battery!)

9. Place a neodymium magnet on the back of the battery and wrap the whole assembly in duct tape, cutting pieces to fit as necessary, so only the glowing LED bulb is showing. Your Throwie dart is done!

---

10. Repeat steps 6 through 9 to make five more Throwies. You should have three Throwie darts of each color.

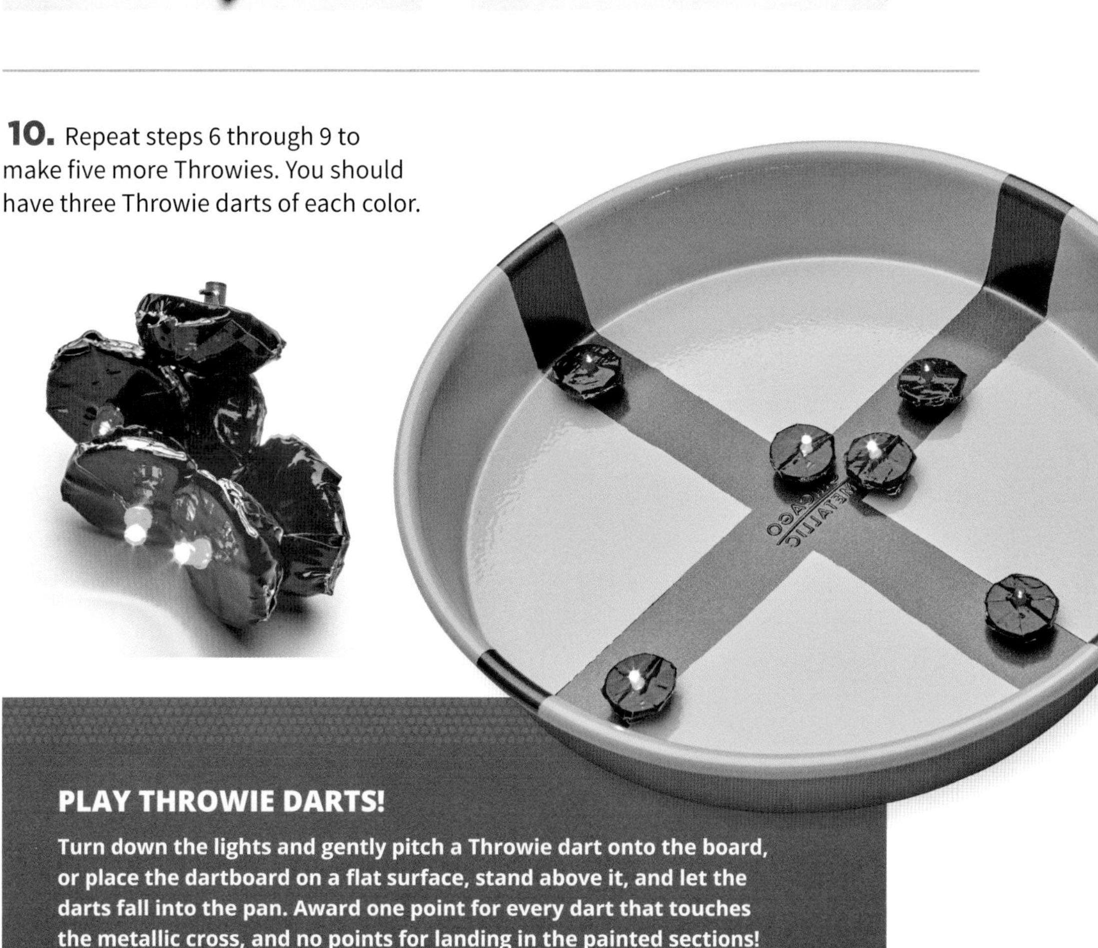

PLAY THROWIE DARTS!

Turn down the lights and gently pitch a Throwie dart onto the board, or place the dartboard on a flat surface, stand above it, and let the darts fall into the pan. Award one point for every dart that touches the metallic cross, and no points for landing in the painted sections!

That's the game we made up. Now make up one of your own.

CHAPTER 2
Paper Circuits

Strange as it may seem, you can make an electrical circuit on a piece of paper. Stick-on LEDs and conductive materials make it possible to create everything from animated greeting cards to toys that twinkle and twirl.

Rather than connecting an LED directly to a battery, as you do when you make a Glowie (page 14), the projects in this chapter separate the battery from the bulb, connecting them instead with tracks of conductive copper tape. When everything is aligned correctly, the copper tape carries the electrical charge from the battery to the LED, lighting it up.

The activities in this chapter use tiny, flat, stick-down LEDs called circuit stickers. These little LEDs may not look like much, but they pack a powerful punch. Like regular LEDs, circuit stickers come in lots of colors and types. Instead of having different-size metal legs sticking out, their polarity (the positive and negative ends) are marked on each sticker: the fat ends are positive (+); the pointy ends are negative (–).

These activities were designed by artist Jie Qi of the MIT Media Lab. Jie is a designer and inventor who gives talks, teaches workshops, and creates artwork with paper circuitry. Her goal is to spread the joys of combining paper craft with circuitry, and encouraging everyone to create their own personally meaningful technology. She is the founder of Chibitronics.

Start by Making a Simple Paper Circuit

SKILL LEVEL 1

For the projects in this chapter, the first step is just to get a feel for what can be done with a paper circuit and how to use circuit stickers in paper-based projects.

Get your tools & materials...

TOOLS

- Pencil
- Craft knife

MATERIALS

- Sheet of paper, any size
- One 3V coin-cell battery (CR2032)
- Copper tape
- One circuit sticker
- Large paper clip or small binder clip

...and MAKE it!

MAKE THE SWITCH

1. Place the battery in one corner of the paper, and then use the pencil to trace around the battery. Move the battery diagonally toward the center of the page, as shown (about ¼" from the first circle), and trace another circle. Mark one circle with a plus (+) sign and the other with a minus (–) sign; it doesn't matter which. Set the battery aside.

2. Fold down the corner of the paper between the circles, so one circle is on top of the other. This will be your switch. Once the paper circuit and battery are assembled, you'll be able to start the flow of electricity by pressing the paper together at this fold.

Your "switch"

MAKE THE CIRCUIT

3. Place the copper tape on the paper to make the circuit. Start the tape in the middle of one drawn circle, and lay a continuous strip of tape that ends in the middle of the other drawn circle.

Place the copper tape in any design or shape you want, so long as it doesn't cross or touch itself and you don't cut it as you go! If you do, you'll break the circuit. Carefully bend or fold the copper tape to make curves or corners.

4. Decide where on the line of tape you want to place your LED. At that point, use your craft knife to cut a small gap of about ¼" into the copper tape. Then bridge the gap with your LED circuit sticker, making sure the positive (+) and negative (–) ends match the orientation of the drawn circles. The fat end of the circuit sticker is always positive (+); the pointy end is always negative (-).

If a length of the copper tape accidentally breaks, firmly press another piece on top of it to continue the circuit.

MAKE IT GLOW!

5. Place the battery in one of the drawn circles, making sure to match the positive (+) and negative (–) sides as marked. Fold the paper and then hold the battery and paper together with a large paper clip or small binder clip.

When the circuit is complete, it will light up your design!

Now that you know how to make a paper circuit, imagine all the different ways you can use it, creating everything from greeting cards to art projects.

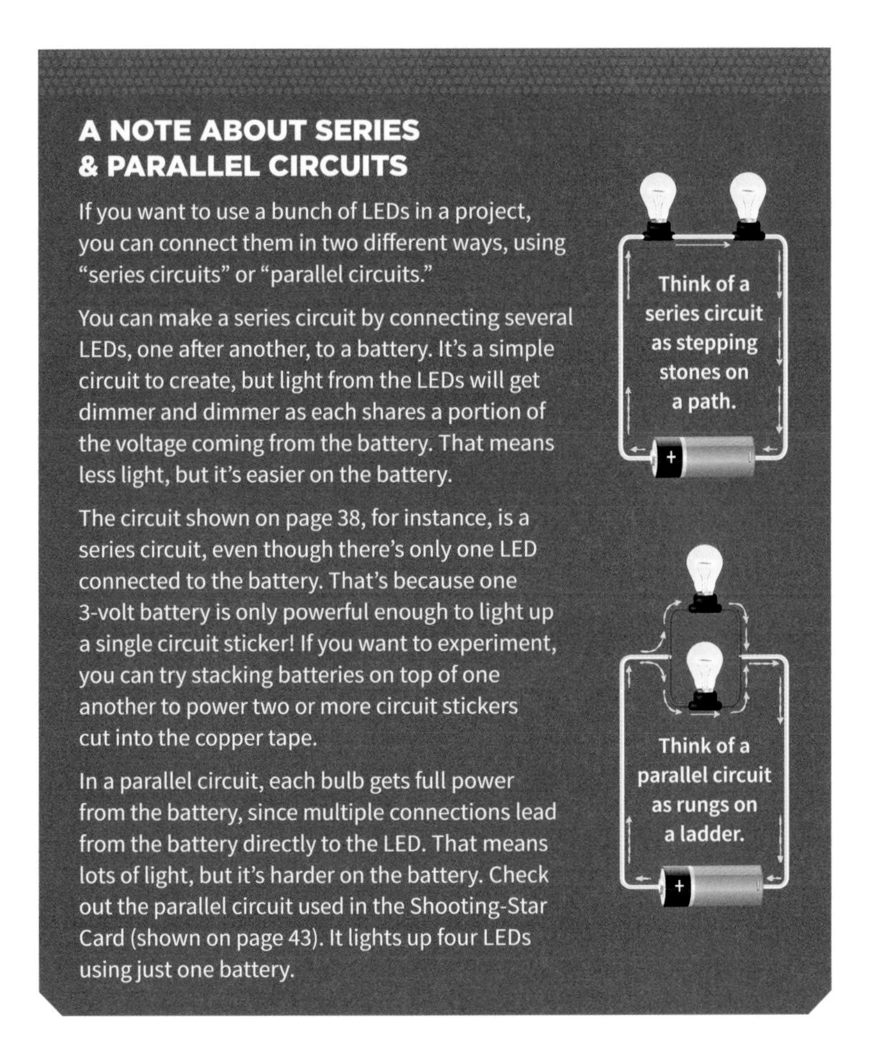

A NOTE ABOUT SERIES & PARALLEL CIRCUITS

If you want to use a bunch of LEDs in a project, you can connect them in two different ways, using "series circuits" or "parallel circuits."

You can make a series circuit by connecting several LEDs, one after another, to a battery. It's a simple circuit to create, but light from the LEDs will get dimmer and dimmer as each shares a portion of the voltage coming from the battery. That means less light, but it's easier on the battery.

The circuit shown on page 38, for instance, is a series circuit, even though there's only one LED connected to the battery. That's because one 3-volt battery is only powerful enough to light up a single circuit sticker! If you want to experiment, you can try stacking batteries on top of one another to power two or more circuit stickers cut into the copper tape.

In a parallel circuit, each bulb gets full power from the battery, since multiple connections lead from the battery directly to the LED. That means lots of light, but it's harder on the battery. Check out the parallel circuit used in the Shooting-Star Card (shown on page 43). It lights up four LEDs using just one battery.

Shooting-Star Card

SKILL LEVEL 1

This clever card animates a city skyline with a "shooting star." Once you get a feel for how to do this, you'll be able to animate other images, as well—balls flying into hoops, food tossed into mouths, UFOs crashing into planets.... You get the idea.

Get your tools & materials...

TOOLS

- Scissors
- Pencil
- Craft knife
- Glue stick

MATERIALS

- Black construction paper
- White or other light-colored blank card
- One 3V coin-cell battery (CR2032)
- Copper tape
- Four LED circuit stickers
- Large paper clip

...and MAKE it!

SET THE SCENE

1. Use the scissors to cut a piece of black construction paper the same size as the front of the blank card.

2. With the pencil, draw a skyline or other silhouetted scene on the bottom third of the black paper. Use the craft knife to cut out the scene (and make sure to cut on a safe surface!). Then, above the cutout, draw four stars in the sky in a line, as shown, and cut them out as well.

Your black paper should now look something like this:

BONUS! Cut carefully and you can use the leftover black paper, as we did, to make another card with stars that shine above a darkened scene.

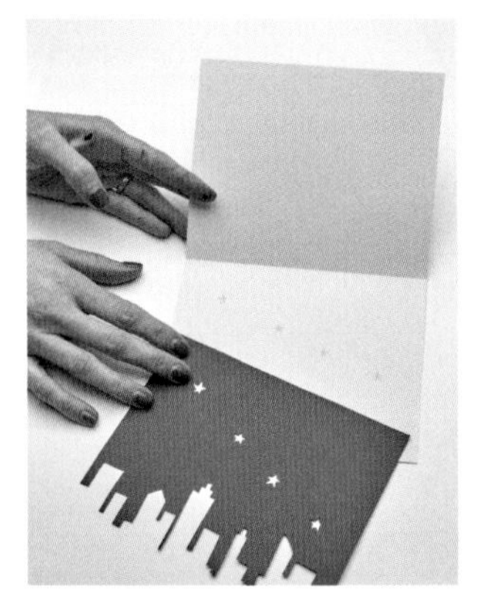

3. Open the blank card and place the black scene inside. With the pencil, lightly mark the position of the four stars. The LEDs will go in those spots.

4. With the glue stick, glue the black silhouette to the front of the card. The outside of your card is now done. The LEDs will shine through the white paper where the stars are cut out.

GLOW IT UP

5. Open to the inside of the card. On the side where the star placements are marked, fold up the bottom inch and trace the battery in the bottom left-hand corner, once above the fold and once below the fold, about ¼" down.

Mark the top circle with a plus (+) sign. The smooth, positive (+) side of the battery will go here. Mark the bottom circle with a negative (–) sign. The bumpy, negative (–) side of the battery will go here.

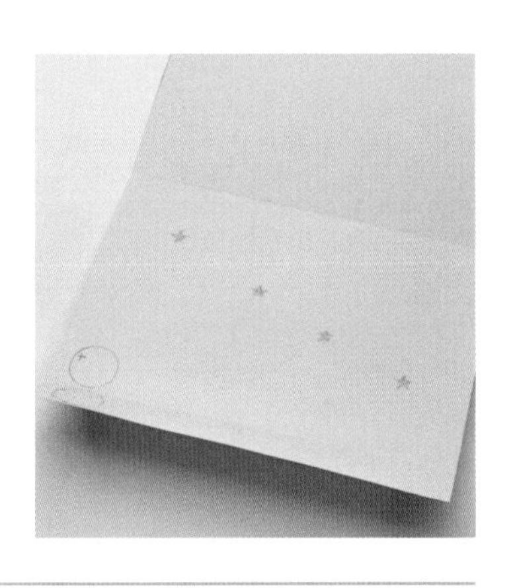

6. Starting from the positive (+) circle, place a line of copper tape till it reaches just above the first star. Then bend the tape (don't cut it!) and run it across to the right, just above the tops of all four stars, as shown. Cut at the end.

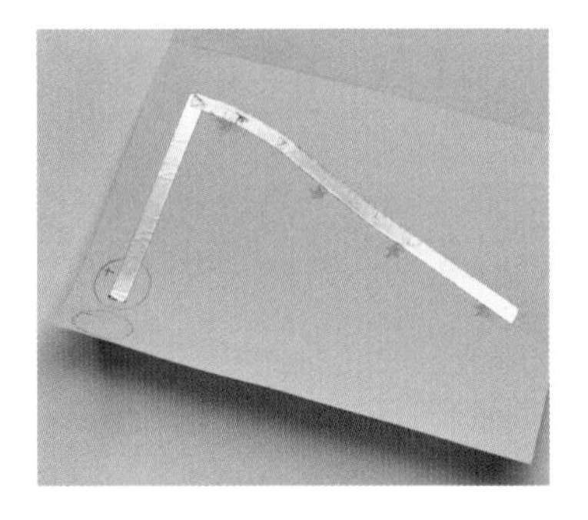

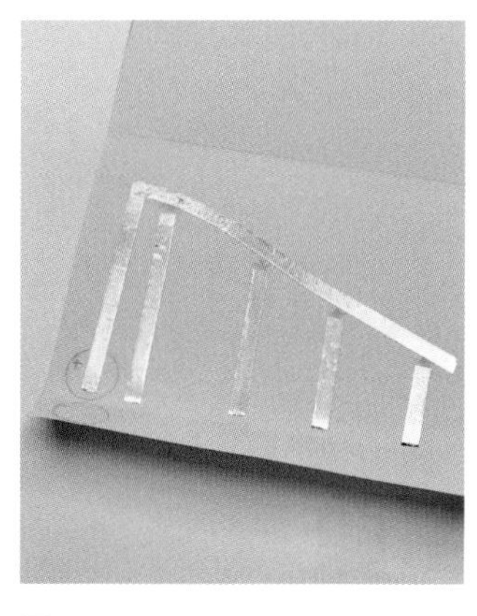

7. Next, stick lines of copper tape from the fold at the bottom of the card to just below each star. Be sure to leave a small gap (at least 1/8" between the horizontal and vertical lines of copper tape.

8. Add a new line of copper tape from the center of the negative (–) circle across the fold-up flap to the end of the card.

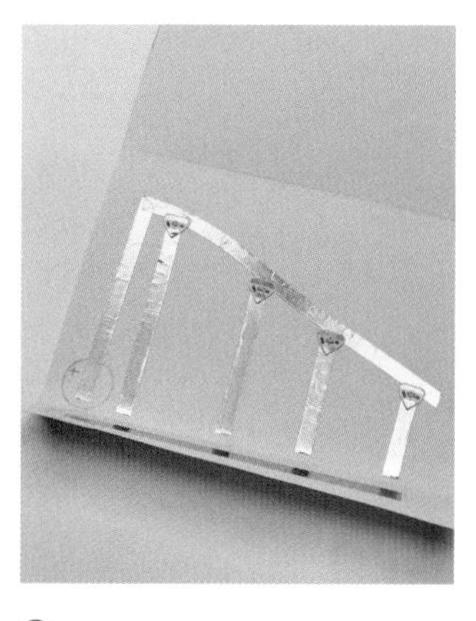

9. Place the four LED circuit stickers, one on each of the star gaps, pointy or negative (–) side down.

10. Place the battery on the positive (+) circle and fold up the flap. Secure with the paper clip and close the card.

If everything's working the way it should, the first star will light up. Then run your finger along the fold-up flap, and the other stars will light one at a time—just like a shooting star in the nighttime sky!

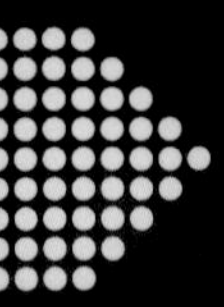

Origami Firefly

SKILL LEVEL 2

How could we resist celebrating nature's own illuminated insect? This classic origami bug is transformed into a flickering firefly with a bit of copper tape and a stick-on LED.

Get your tools & materials...

TOOLS

- Pencil
- Scissors

MATERIALS

- Origami paper (or any square piece of foldable paper)
- Transparent tape
- Piece of cardstock or index card, 1" x 2"
- One 3V coin-cell battery (CR2032)
- Two small neodymium button magnets
- Copper tape
- One LED circuit sticker
- Optional: googly eyes, colored markers, and so on (for decorating your firefly)

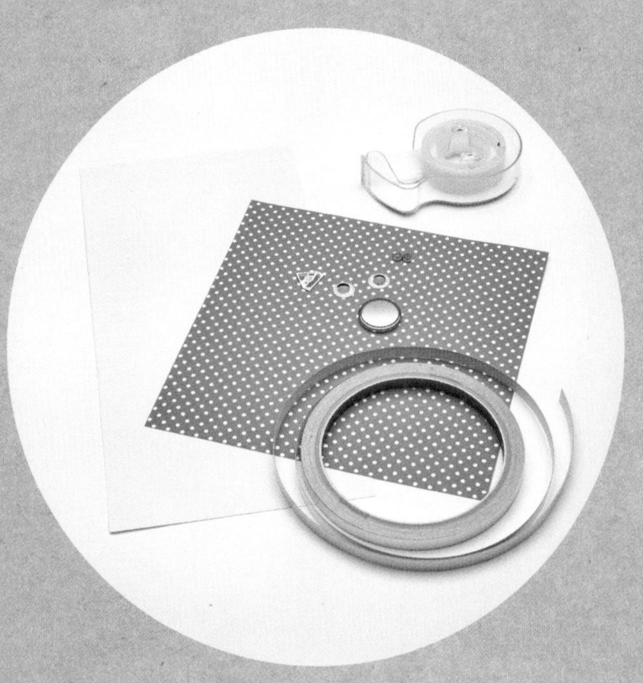

...and MAKE it!

MAKE YOUR FIREFLY

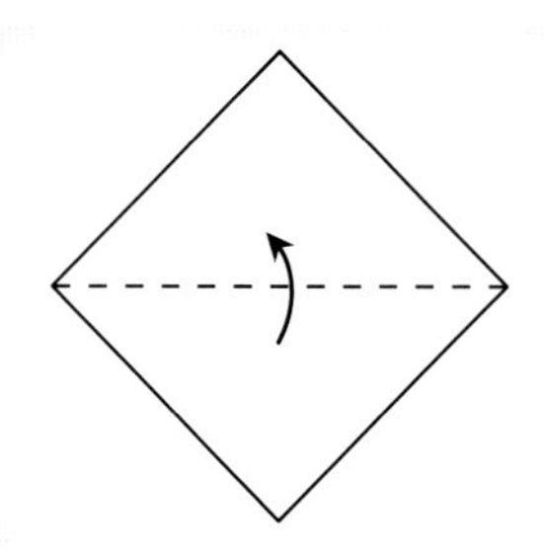

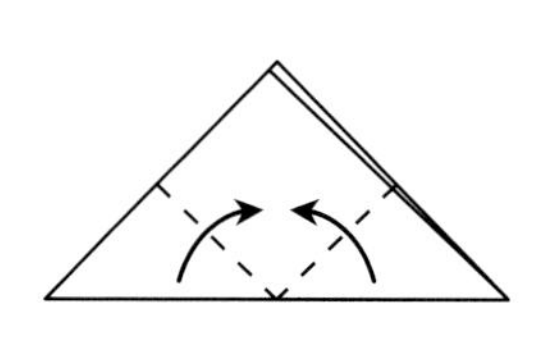

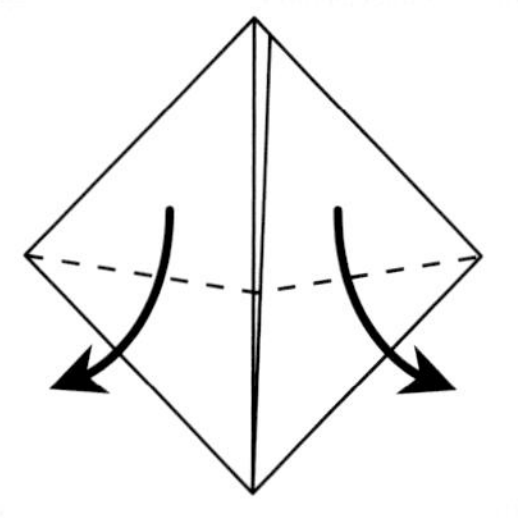

1. Place the square sheet of paper in front of you, oriented like a diamond. Fold up the bottom half so the peaks meet at the top.

2. Fold up the two bottom corners so they meet at the top.

Then fold the same two flaps down, folding them in the middle, so the corners point away from each other. (They're the wings of your firefly.)

3. Pick up the top layer at the unfolded tip (above the wings) and fold down, so the bug's "butt" is sticking out a bit.

Then fold down the bottom layer of the tip to about ¼" from the middle layer.

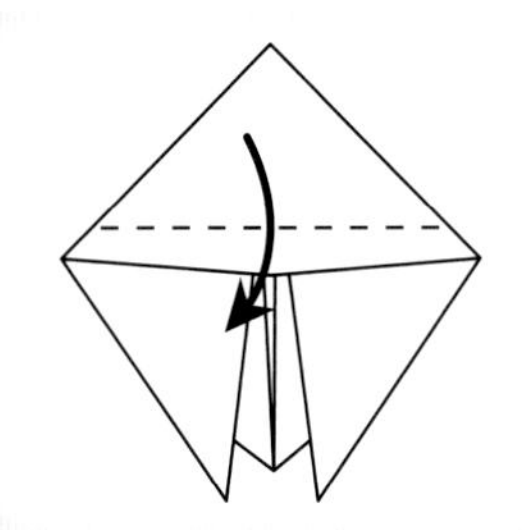

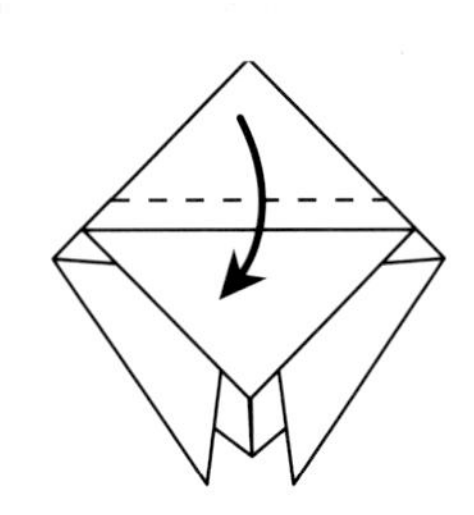

4. Finally, bend the two sides to the back and tape in place, as shown.

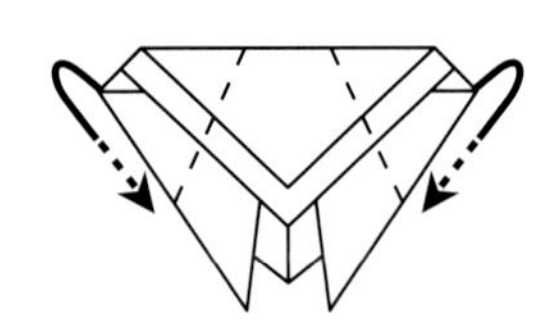

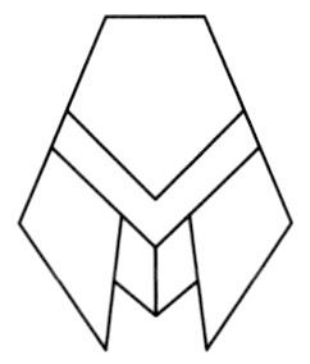

Your origami firefly is ready.

ATTACH THE BATTERY HOLDER

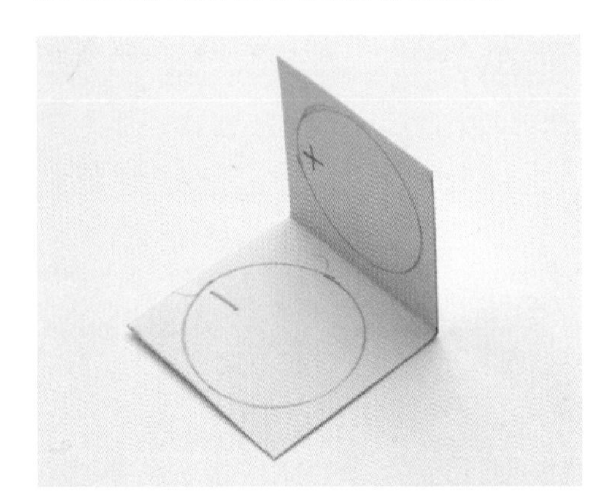

5. Fold the 1" x 2" piece of cardstock in half.

Flatten it out. With the pencil, trace around the battery on each half, as shown. Mark one circle with a plus (+) sign and the other with a negative (–) sign.

6. Tape one button magnet to the center of each circle, and then tape one side of this battery holder to the back of the firefly. As you do, make sure the magnets attract (rather than repel) each other, and make sure the negative (–) side of the battery holder is attached to origami firefly.

CREATE THE CIRCUIT

7. On the negative (–) side of the battery holder, place a piece of copper tape that starts at a point just above the magnet and runs the length of the bug. Cut it at the bottom. Then, wrap another small piece of copper tape around the end of the bug's butt that connects to the first piece of tape. (If you can bend the copper tape around the bug's butt without cutting it, that's even better!)

8. On the positive (+) side of the battery holder, place a piece of copper tape that starts just beyond the magnet and goes diagonally to the outer edge of the holder, down the length of the bug, wraps around its butt, and comes back up on the other side, as shown.

Cut the tape just below the circle for the negative (–) side of the battery holder.

9. Flip over the bug and add the LED sticker, pointed side down, so that it connects the two strips of copper tape, as shown.

10. Decorate the firefly by drawing a face, sticking on googly eyes, or anything else you'd like.

When you're done, pop the battery in the holder (the magnets will keep it in place), and your firefly will come alive!

Glowing Pinwheel

SKILL LEVEL 3

A DIY pinwheel is fun anytime, but this one creates great patterns of spinning, sparkling light.

Get your tools & materials...

TOOLS

- Pencil
- Ruler
- Scissors
- Hot-glue gun and glue sticks

MATERIALS

- One 6" x 6" square of origami paper, construction paper, or similar
- Copper tape
- Cardstock scraps
- One 3V coin-cell battery (CR2032)
- Transparent tape
- Three small neodymium magnets
- Pin clasp with back
- Four circuit-sticker LEDs (your choice)
- Sturdy plastic straw

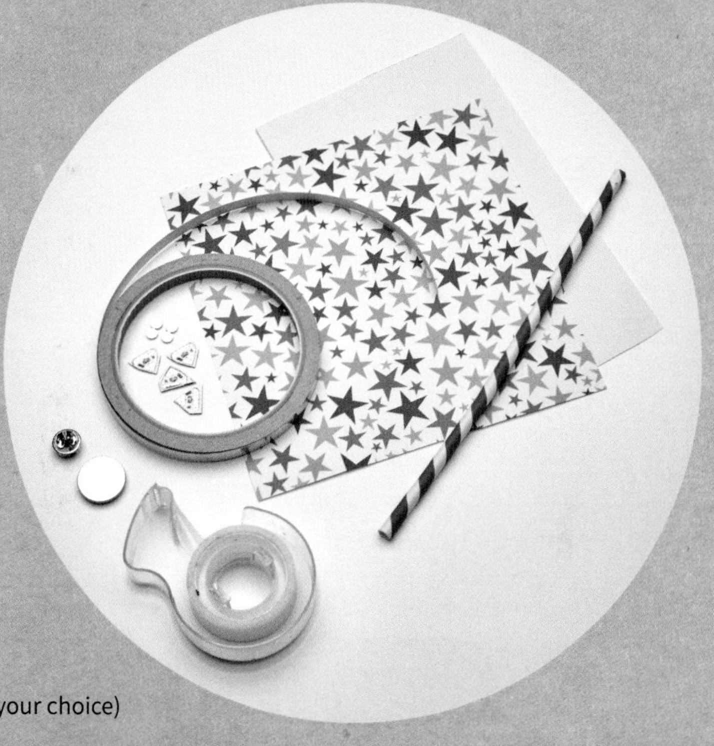

...and MAKE it!

PREP THE PINWHEEL

1. If your paper is decorated, flip it over to work on the plain side.

With the pencil and ruler, draw two diagonal lines from corner to corner. Then measure three inches down from each corner and mark that point on each line. Using scissors, cut along each diagonal line down to each mark. Going clockwise, label each pinwheel section lightly with a pencil as A, B, C, and D, as shown. You can erase the letters later.

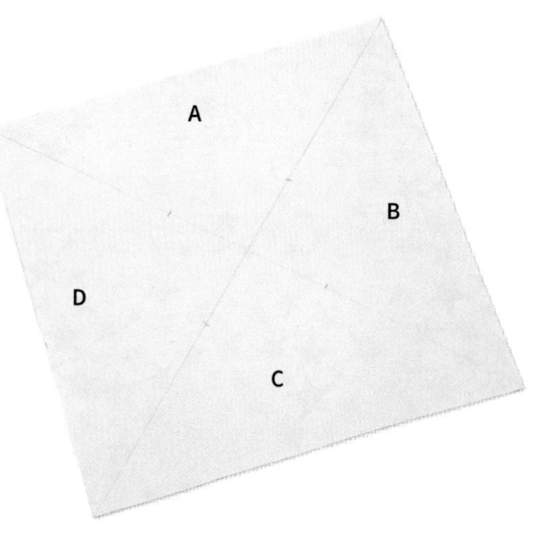

SET UP THE CIRCUIT

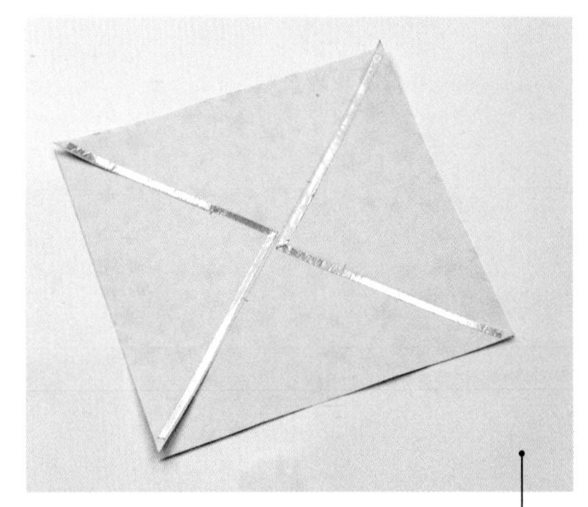

2. Place a line of copper tape from the outside edge of section A to the outside edge of section D. Bend the tape in the center of the pinwheel, since the tape needs to be in one piece. Snip at the end. Repeat by placing a second line of copper tape along the outside edges of sections B and C. Don't let the two lines of tape touch in the center of the pinwheel. Your paper should look like this:

---

3. Using scissors, cut three scraps of cardstock, each about ¼" x ½" These serve as "bridges" to keep crossing copper lines from touching (which would short out the circuit).

Using the hot-glue gun, glue one cardstock bridge on the D line near the center of the pinwheel, and glue a second cardstock bridge on the B line near the center of the pinwheel, as shown. (It's okay to glue on top of the copper tape.) The bridges should completely cover the copper tape lines.

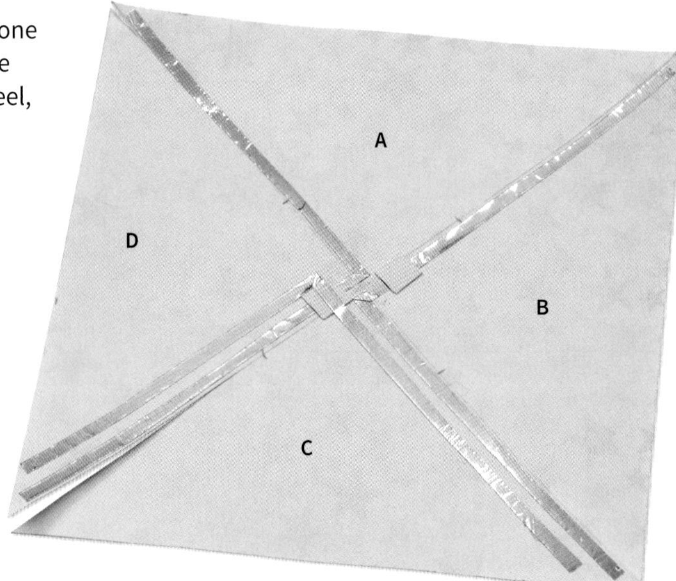

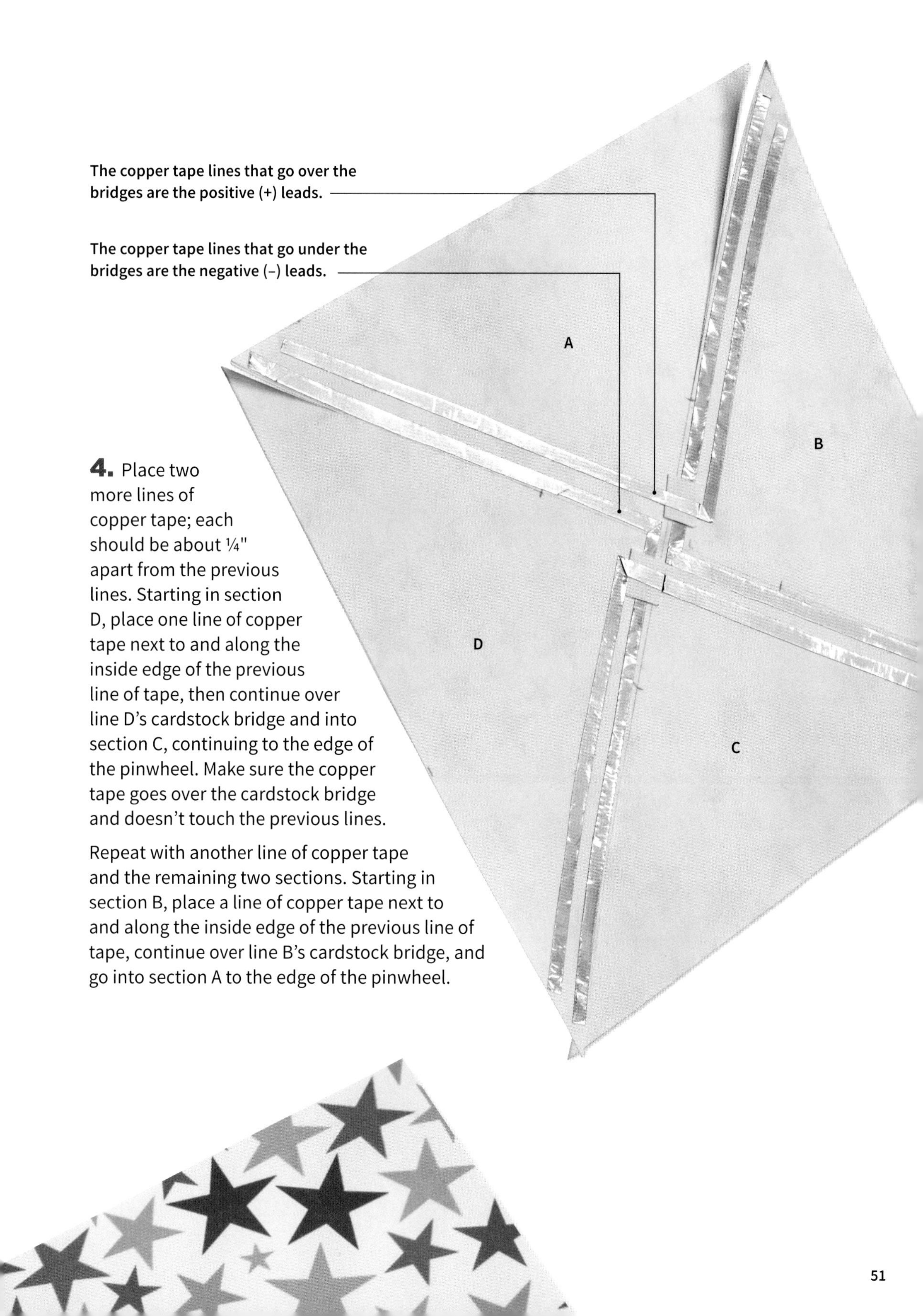

4. Place two more lines of copper tape; each should be about ¼" apart from the previous lines. Starting in section D, place one line of copper tape next to and along the inside edge of the previous line of tape, then continue over line D's cardstock bridge and into section C, continuing to the edge of the pinwheel. Make sure the copper tape goes over the cardstock bridge and doesn't touch the previous lines.

Repeat with another line of copper tape and the remaining two sections. Starting in section B, place a line of copper tape next to and along the inside edge of the previous line of tape, continue over line B's cardstock bridge, and go into section A to the edge of the pinwheel.

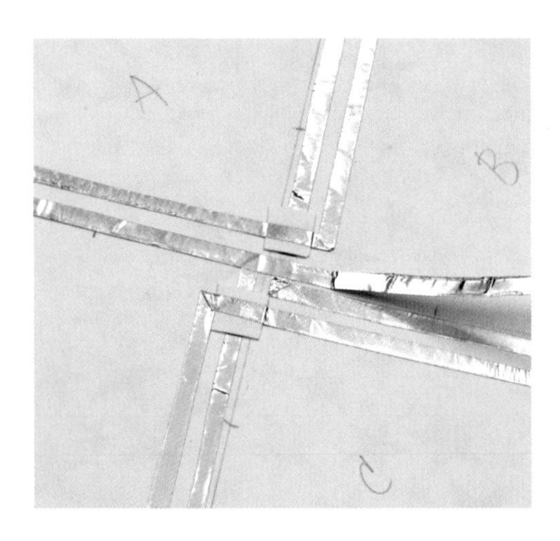

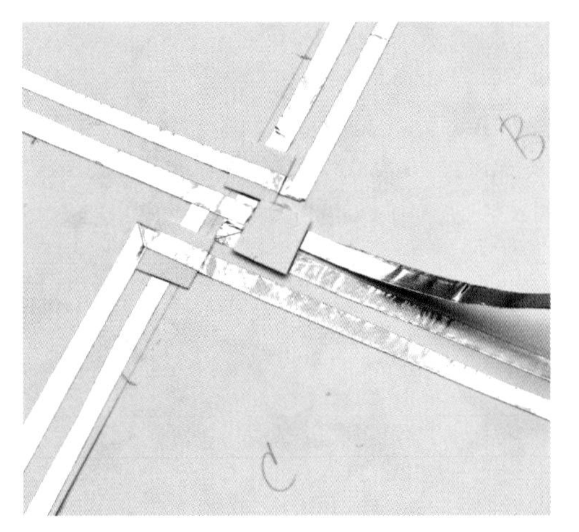

5. Using scissors, cut a piece of copper tape 8" long. Lay about 2" of it across the center of the pinwheel, so that it lays right on top of the negative (–) lead in section A and connects to the negative (–) lead in section B (the negative leads are the lines running under the bridges). With the remaining 6" of copper tape, fold it over on itself to make a 3" tab. Remember to stick the copper tape down tight to make a good connection.

Using the hot-glue gun, glue the third cardstock bridge at the base of the 3" tab so that the bridge also covers the adjacent section of negative (–) lead near the base of section C, as shown. The bridge must cover both the tab and the negative (–) lead.

---

6. Using scissors, cut another 8" strip of copper tape. Lay about 2" of it across the center of the pinwheel, so it sits right on top of the positive (+) lead in section B, crosses over the bridge you placed in step 5, and connects to the positive (+) lead in section C (the positive leads are the lines that run over the bridges). With the remaining 6" of copper tape, fold it over on itself to make a 3" tab. As always, make sure the copper tape is stuck down tight!

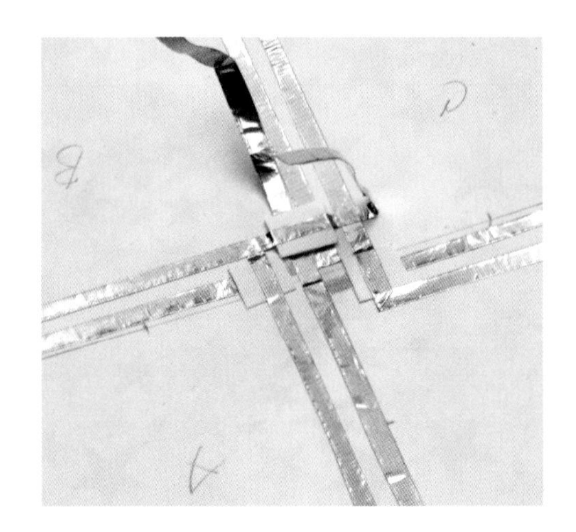

ADD THE BATTERY HOLDER

7. Using scissors, cut a piece of cardstock about 1" x 3" and fold it in half. Trace the battery on each side, marking one circle with a plus (+) sign and the other with a minus (–) sign, as shown. Using transparent tape, tape a neodymium magnet into the center of each circle (make sure the magnets attract!). The magnets will keep the battery in place, but allow you to remove it whenever you want.

---

8. Set the battery holder in the center of the pinwheel, negative (–) side down. Fold the copper tab from the negative (–) lead onto the negative (–) side of the battery holder, as shown (trim the tab as needed). Secure the tab with a small piece of copper tape. Then use transparent tape to secure the negative (–) side of the battery holder to the pinwheel.

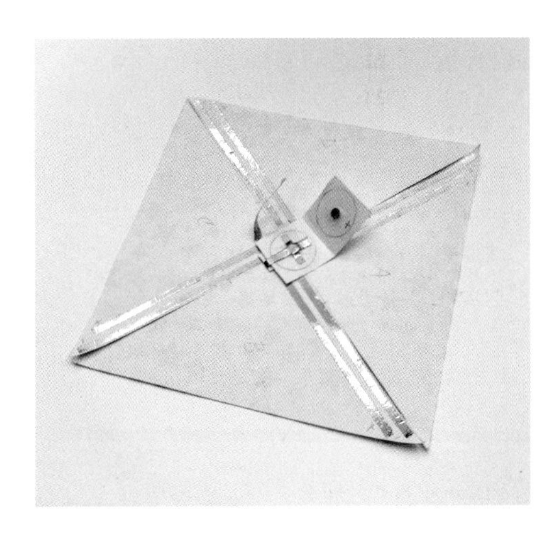

ASSEMBLE YOUR PINWHEEL

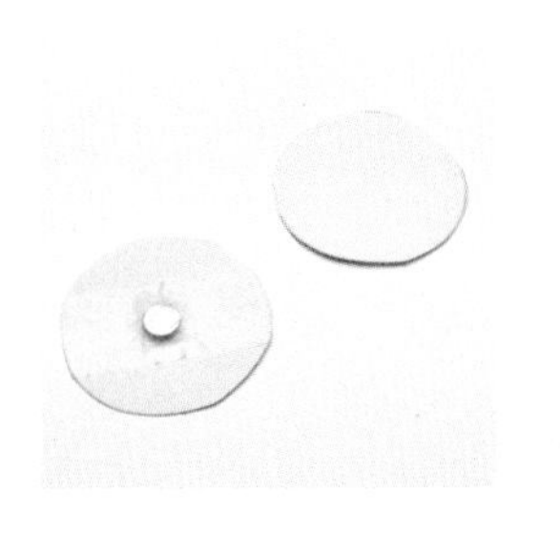

9. Using scissors, cut three cardstock circles, each about 1" in diameter. Using transparent tape, tape the third magnet to one circle, stick the pin of the pin clasp through the second (save the pin back), and leave the third plain.

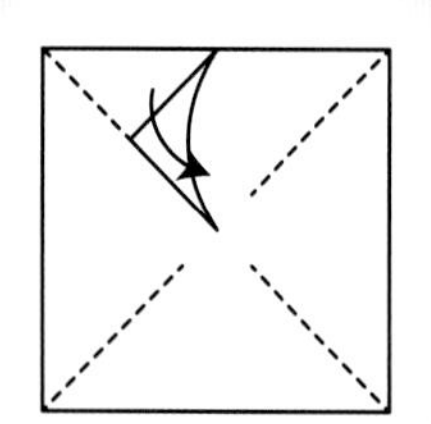
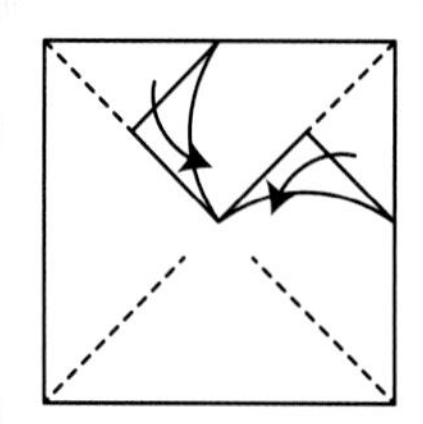
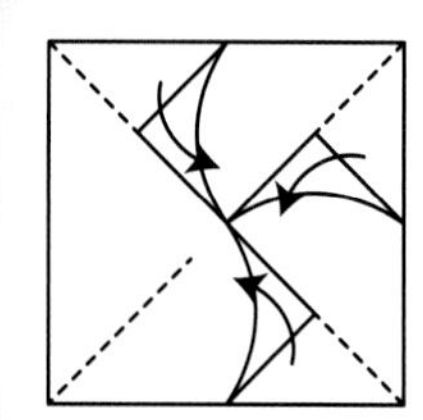
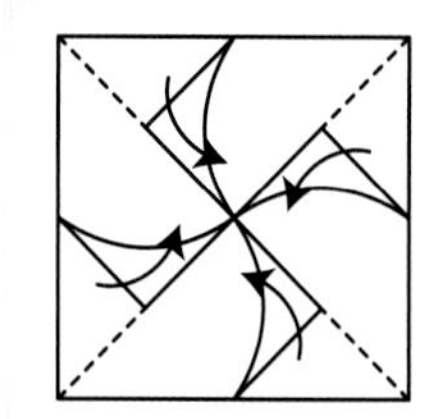

10. Test the circle with the magnet to make sure the magnet attracts to the battery holder by placing the magnet side to the outside of the holder. If it doesn't attract, reverse the battery on the circle. Then gently bend each unused corner of the pinwheel toward the center of the pinwheel, and with the hot-glue gun, glue each corner to the back of the circle with the magnet. Once the glue is dry, glue the plain circle on top to hide the assembly. Once done, the magnets will hold the pinwheel together, while still allowing access to the battery.

ADD LEDS ... AND GLOW!

11. Once the pinwheel is assembled, place one circuit-sticker LED on each pinwheel blade so it connects the two copper-tape leads. The pointy, negative (–) side of each sticker goes on the outside edge of each blade; the fat, positive (+) side of each sticker goes on the inner line of copper tape.

TEST IT! Pop in the battery and make sure all the LEDs light up!

12. Finally, using the hot-glue gun, glue the circle with the pin to the back of the pinwheel in the center. Poke the pin through the straw, and secure it with the pin back.

Et voila!

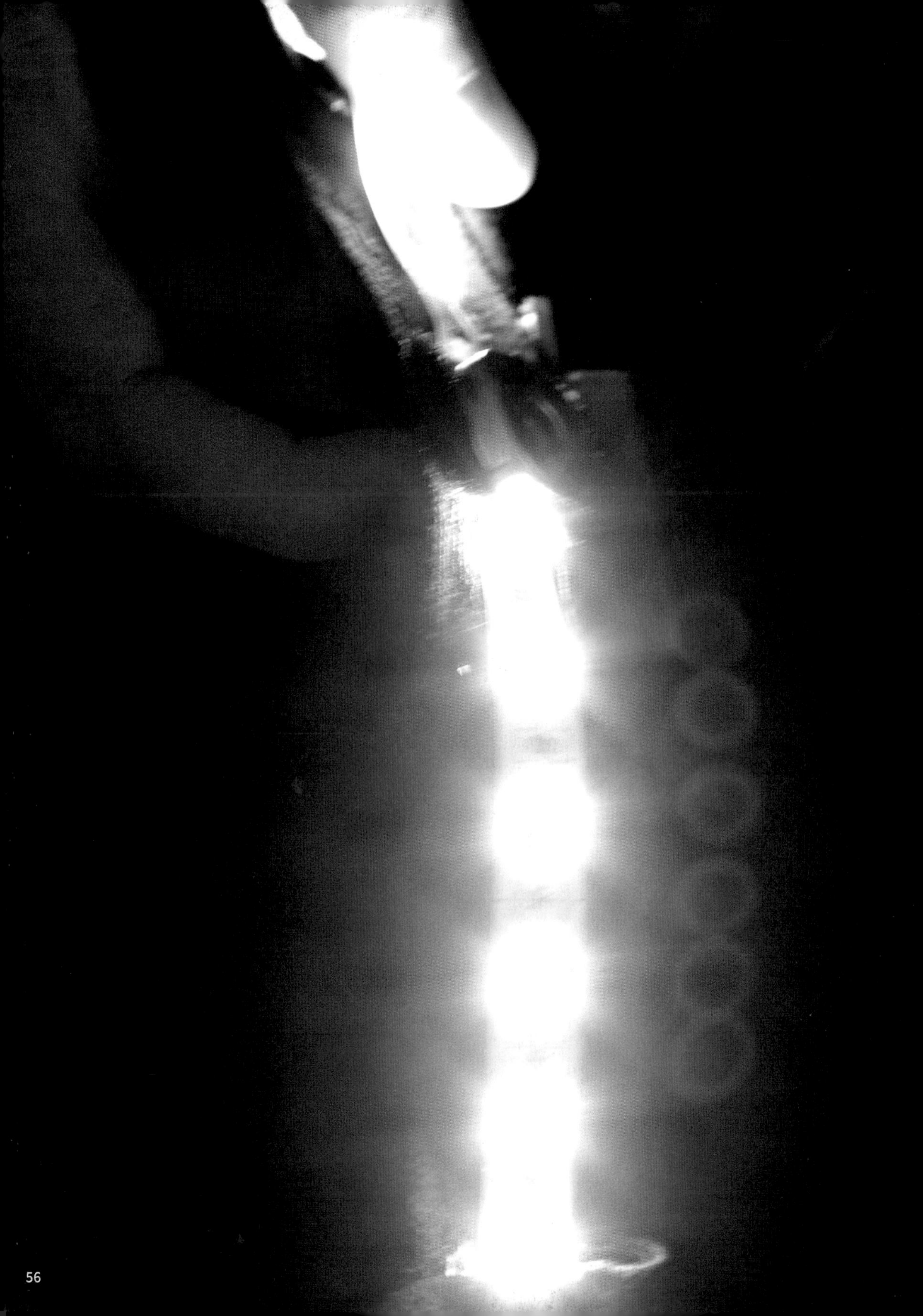

CHAPTER 3
Soft Circuitry & Wearables

Power up your wardrobe with activities that you (and even your dog) can wear!

This chapter introduces everything you need to glam up your wardrobe with sewable LEDs, switches, battery packs, and more. Stitch them all together with metallic conductive thread, and you can turn clothing and accessories into one-of-a-kind artworks that show off your style and ingenuity—even in the dark.

This section also introduces a few new tools. The third-hand-tool—bristling with clips and clamps so you can hold components steady—is inexpensive and, well, handy. Some even come with magnifiers to help you work on even the tiniest bits and pieces.

You'll also need wire strippers for many of these projects. Most wire strippers have wire-cutting capabilities, but if yours don't, you'll need wire cutters as well.

Also new in this chapter is the use of resistors. Using resistors will protect your LEDs from burning out and help your batteries last longer.

More challenging is the soldering iron. Get an adult to help you solder. Don't mess around with this tool! Hot irons and molten metal are dangerous. But when used with care and respect, soldering is a skill that offers new opportunities for creativity.

Power Cuff

SKILL LEVEL 2

Feel the power when you rock this glimmering cuff. For this project, a friend can be helpful when it comes time to measure.

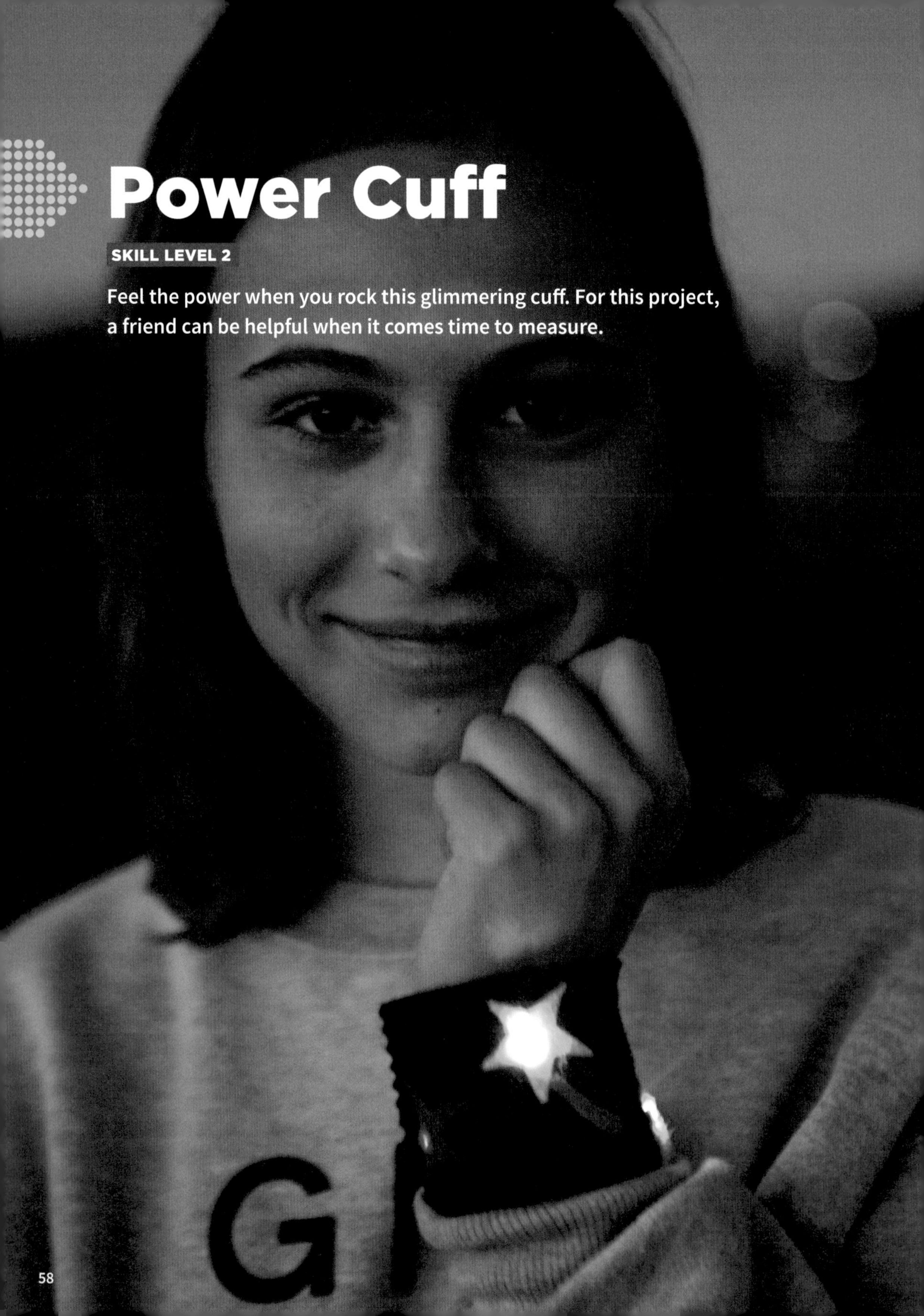

Get your tools & materials...

TOOLS

- Scissors
- Black marker
- Chalk
- Needle-nose pliers
- Hot-glue gun and glue sticks
- Sewing needle
- Optional: pinking shears

MATERIALS

- One 8.5" x 11" piece of black felt
- Scrap felt, various colors
- Five 2mm diffused flat-top white LEDs
- Sewing thread
- Two sew-on snaps
- Sew-on battery holder
- Conductive thread
- One 3V coin-cell battery (CR2032)
- Clear nail polish

...and MAKE it!

DESIGN YOUR CUFF

1. Using scissors, cut the 8.5" x 11" piece of black felt in half lengthwise to make two 4.25" x 11" strips. Set one aside to use as scrap in step 8.

Cut the other piece in half lengthwise, so you have two thin strips about 2.1" x 11" long. Wrap both around your wrist to measure your wrist size, add 1" to this length, and cut both pieces. For now, set both aside. One piece will become your cuff, and the other you may want to use as a cuff liner.

2. With the black marker, draw any shapes you want on the scrap pieces of colored felt. Our design uses four stationary stars and one falling star with three "swoop" marks. Other design ideas are a sun and crescent moon, planets with rings, puppy faces with light-up eyes—anything you want. Whatever you design, leave at least an inch blank at both ends of the felt (so you can add the snaps).

We used five LEDs, which made a sparkly cuff without putting undue stress on the battery. You can use fewer LEDs, but we don't recommend trying to use more.

3. When the design is finalized, use scissors to cut out the shapes. Set them in place to test how they fit—but don't glue them down yet: The LEDS will go under the felt and shine through. Figure out where you want each LED to be, use chalk to mark each spot on the black felt, and then set your felt design pieces aside. You'll add them in step 7.

SET YOUR LEDS AND DESIGN IN PLACE

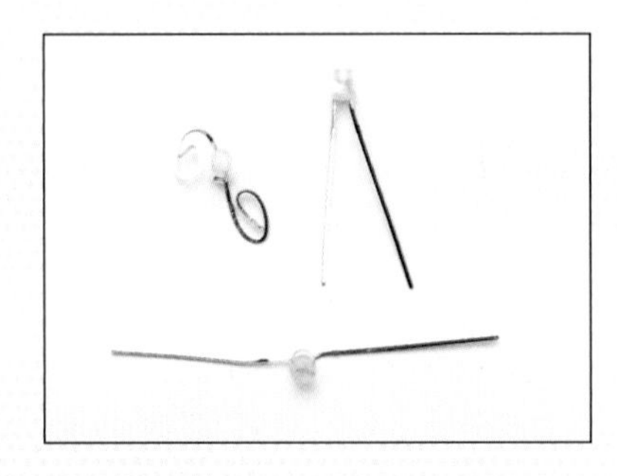

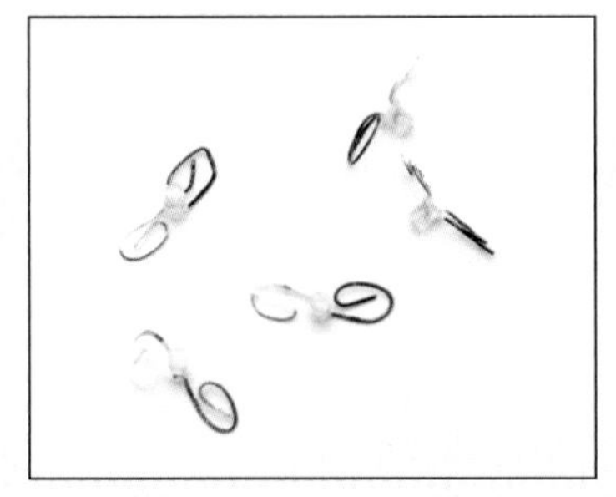

Bend the wire gently, so you don't break off the LED legs.

4. With the black marker, darken the long positive (+) legs of each LED. Gently bend the legs in opposite directions to flatten each LED, and then use the needle-nose pliers to twist each side into a curl.

5. Using scissors, snip small holes in the felt at each chalk mark and insert the coiled LEDs from the back of the cuff to the front, so that only the LED light appears on the front, and both coiled legs are in back. Darken the positive (+) leg of each LED with more marker if necessary, so you can always see them.

6. Working on the back, carefully arrange the coiled LED legs so they're aligned for easy sewing. We've turned ours so all the positive (+) legs are to the inside of the cuff, and all the negative (–) legs are to the outside, as shown. This kind of arrangement simplifies the conductive-thread sewing in steps 10 to 14.

7. When all the LEDs are in place, use the hot-glue gun and, working from the front, put a dab of hot glue on top of each LED light. Then add all the cutout felt pieces that make up your design.

SNAP IT TOGETHER

8. On the back of the cuff, about half an inch from one end, use a needle and regular thread to sew on the male sides of the two snaps. (Remember: Knot the end of the thread each time you sew.)

Then, wrap the cuff around your wrist and mark where the female sides of the snaps will need to go on the other end. This is easier if a friend helps you hold the cuff and measure. Using regular thread, sew the female sides of the snaps in place.

---

9. If you want fancy edges, use the pinking shears to trim the ends of the cuff, or use scissors to trim any extra felt overhang.

Take the set-aside, second 4.25" x 11" strip of black felt. Using either scissors or pinking shears, cut a piece that's a bit wider than your battery holder and the same height as the cuff. Using the hot-glue gun, glue it to the front of your cuff where the male sides of the snaps are sewn, and then glue the sew-on battery holder on top. Now you're ready to sew the circuit!

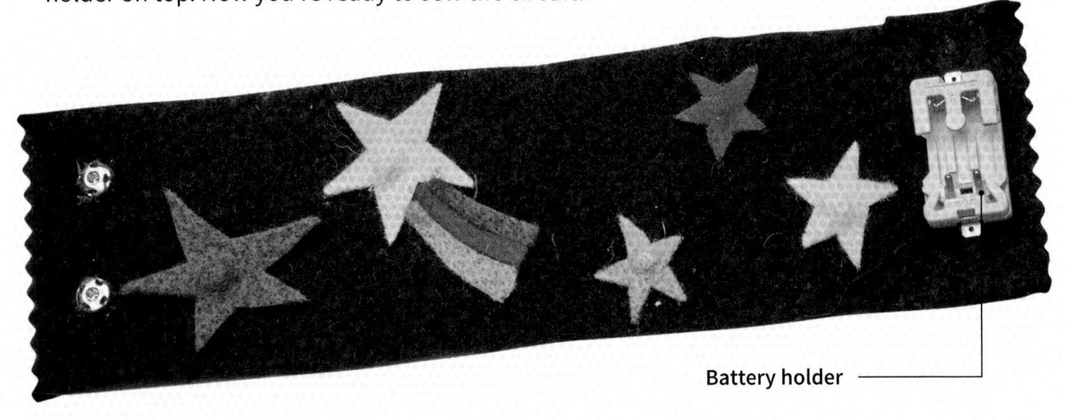

SEW AND SHINE!

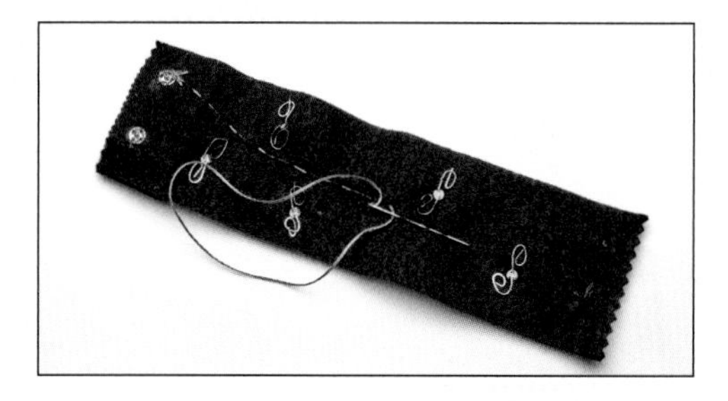

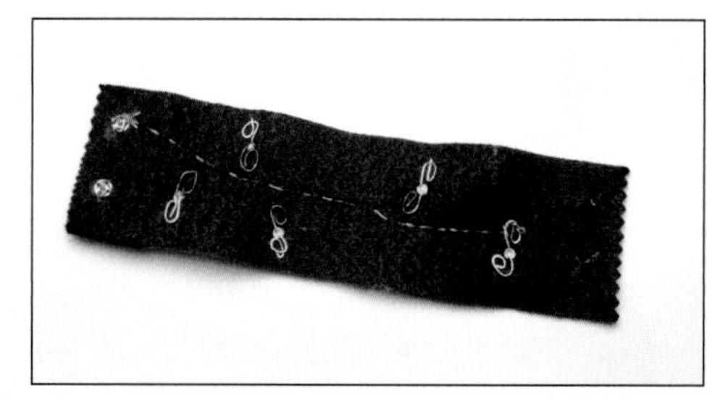

10. Using a needle and conductive thread, start from the back and push the needle to the front, going through the sewing hole at the positive end of the battery holder, and return the needle to the back.

Continue sewing on the back. Thread through the surface layers of the felt—without touching any of the metal legs of the LEDs!—until you get to the LED that's farthest away. Sew around the curled positive (+) leg of the farthest LED. Knot and clip the thread.

This is the positive (+) "bus" lead, providing power from one end of the cuff to the other. You can think of it like a real bus, offering service to every stop along its route.

11. Using a needle and conductive thread, attach each positive (+) leg of the four remaining LEDs to the positive (+) bus lead. Sew the first leg to the lead, then knot and clip the conductive thread. Then do the same for the remaining LEDs. Each should be connected individually.

12. Now create the negative (–) side of the circuit. Using a needle and conductive thread, sew through the male snap under the battery holder to the sewing hole on the negative (–) side of the battery holder, and return the thread to the back. Knot and clip the thread.

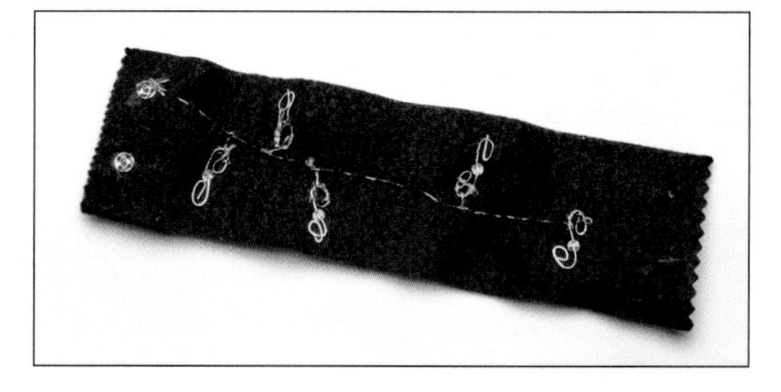

By sewing the snaps into the circuit, you create a simple switch. When you snap on the cuff, you complete the circuit and turn on the LEDs!

13. Match the female snap on the other side of the cuff that lines up with the negative (–) end of the battery holder. Ours is on the bottom, as shown.

Using a needle and conductive thread, sew through that snap several times (start from the back to hide the knot), and finish on the back of the cuff. Knot and clip the thread.

14. Using a needle and conductive thread, and working on the back, sew from the female snap you just connected through the curled-up negative (–) legs of all five LEDs.

Make sure the negative (–) line doesn't touch the positive (+) line, or you'll short out the circuit! If you need to cross conductive threads, glue a felt "bridge" over the first line (using scrap felt), then sew the second line across it, as shown. This is just like the "bridges" used in Glowing Pinwheel, page 48.

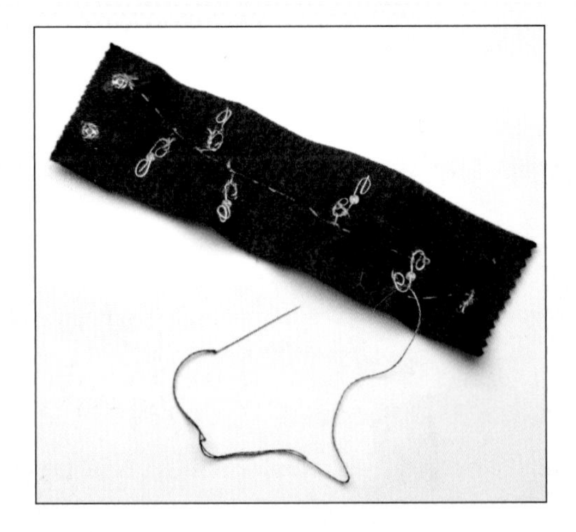

15. For the last LED, knot and cut your thread. You're done! Slip a battery into the holder and snap on your cuff. Since the snap acts as your switch, the LEDs should light up immediately.

When all the sewing is finished and you confirm the circuit works, cover the thread on the back with clear nail polish to keep it secure.

If you want, you can make the cuff more comfortable by sewing or gluing the other 2.1" x 11" wrist-sized strip of felt to the back of the cuff.

Show your glow!

Fashion for Fido: Glow-in-the-Dark Dog Collar

SKILL LEVEL 3

Luminous LEDs glam up your pet's collar. What pampered pup could ask for more?

Get your tools & materials...

TOOLS

- Craft knife
- Hot-glue gun and glue sticks
- Super glue
- Pencil or chalk
- Ruler
- Tweezers
- Wire strippers
- Sewing needle
- Scissors
- Optional: pliers

MATERIALS

- Purchased woven pet collar with clip-style buckle, at least ¾" wide
- One tactile switch with leads
- One battery holder
- Five sew-on LED sequins
- Conductive thread
- One 3V coin-cell battery (CR2032)
- Clear nail polish
- Scrap felt in a coordinating color

...and MAKE it!

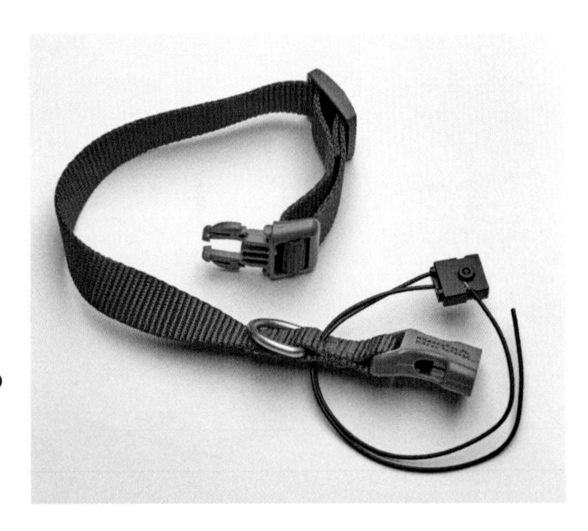

GET THE COLLAR READY

1. Use a craft knife to cut a small hole in the webbing between the D ring and the female side of the buckle on the pet collar. Feed the leads of the tactile switch through the hole so the wires emerge on the *inside* of the collar.

2. Using a hot-glue gun, glue the base of the tactile switch to the outside of the collar, over the hole you just made, as shown. Then, using super glue, secure the webbing around any cut edges, both inside the collar and out, to keep them from fraying.

3. On the outside of the collar, use the hot-glue gun to attach the battery holder on the other side of the D ring from the tactile switch, as shown. Make sure the positive (+) end of the battery holder is facing the buckle and switch.

4. Put the collar on your pet and adjust the length to fit. Remove the collar, and on the outside, use pencil or chalk to mark the places where you want LEDs. (If your pooch is still growing, be sure to add some space so the LEDs will look good even if you have to adjust the collar!) To space the LEDs evenly, start from the center of the collar and use a ruler to measure exactly.

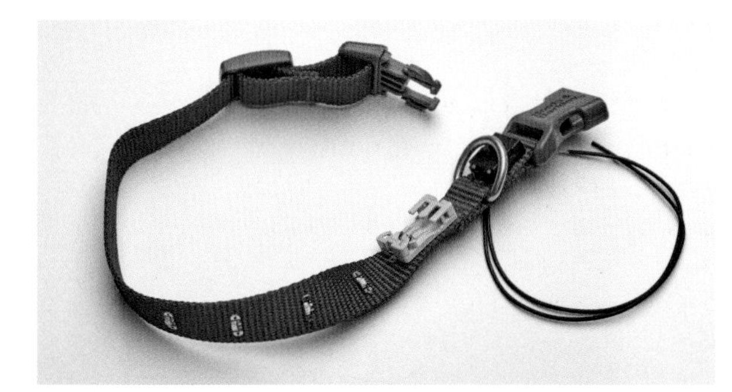

5. When placements are marked, set the collar down flat in front of you with the D ring to the right. Use tweezers to pick up the tiny LED sequins, and use a small dot of super glue to set each one in place. To make sure all the LEDs complete the circuit and are oriented in the same way, glue each so its negative (–) end points toward the top edge of the collar.

6. On the tactile switch, use wire strippers to trim the two leads sticking through on the inside of the collar to about 1" long. Then, strip the plastic off the ends of the leads, exposing about half an inch of the wire inside. If you need, use pliers to help hold everything steady; if the wires separate once exposed, just twist them back together again.

When you're done, use the hot-glue gun to secure the base of the wires to the collar, leaving the exposed wire ends free.

SEW IT UP!

7. Use a needle with conductive thread to sew down the free end of one of the wires you just stripped (it doesn't matter which one you choose; tactile switches have no polarity). Sew around the wire as many times as necessary till it's completely covered. Then push the needle through to the outside of the collar and sew through the sewing hole on the positive (+) side on your battery holder. Knot the thread and cut it.

Don't rush! Conductive thread can easily get twisted and knotted. Take the time to unknot the thread if you start to have trouble.

8. Using a needle and conductive thread, sew through the sewing hole on the negative (–) side of the battery holder (unlike tactile switches, battery holders do have polarity!), and then sew along the surface of the webbing to the negative (–) hole in the first LED. You don't need to push the needle through the collar—just pick up the top threads of the webbing as you sew.

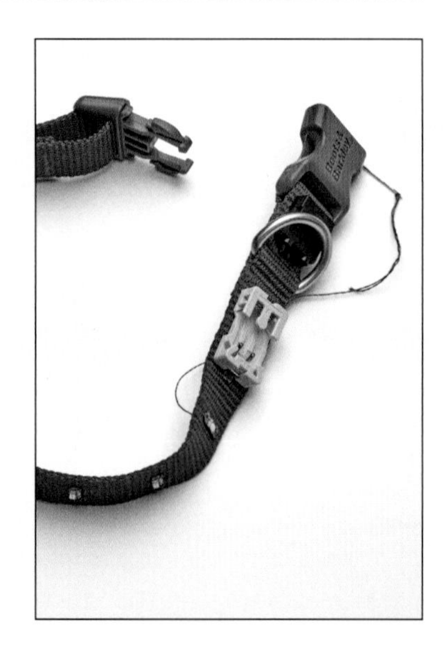

9. Continue sewing until the thread has connected all five LEDs on the negative (–) sides only. Make sure not to cut the thread in between them! And remember, the super glue holds the LEDs in place; the thread just connects them to the circuit. When finished, knot and cut the thread.

10. Using a needle and conductive thread, and starting with the LED where you just ended, sew the LED's positive (+) side, and keep sewing to connect the positive (+) ends of all five LEDs.

11. When the positive (+) sides of all five LEDs are sewn together, and without cutting the thread, push the needle through to the other side of the collar and sew the conductive thread to the second, unattached wire lead of the tactile switch. As in step 7, sew around the wire as many times as necessary to completely cover it with thread, then knot and cut the thread.

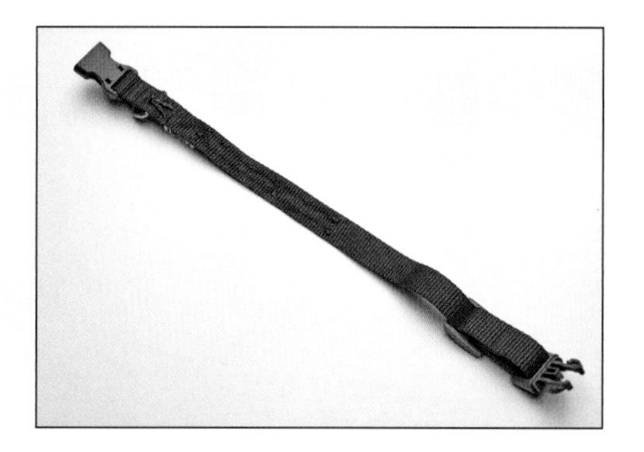

12. When everything's working, cover your sewing with clear nail polish to set it in place and cover any prickly thread. Finish up by measuring and cutting a scrap of felt in a coordinating color to place on the interior of the collar for your pet's comfort. Use the hot-glue gun to attach the felt.

Your pup's gonna love it!

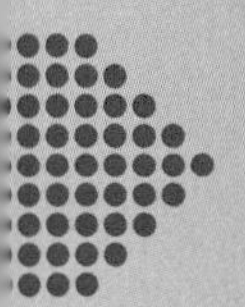

I ❤ My Stuffie

SKILL LEVEL 3

You know your stuffies love you—almost as much as you love them. Here's a way to show it! A flashing LED makes a beating heart on the inside, and a few stitches with embroidery thread shows the love on the outside.

Get your tools & materials...

TOOLS

- Scissors (sewing scissors work best, if you have them)
- Craft knife
- Black marker
- Wire strippers
- Soldering iron and solder
- Third-hand tool
- Needle-nose pliers
- Hot-glue gun and glue sticks
- Pencil or chalk
- Sewing needles

MATERIALS

- Hackable stuffie
- Tactile on/off switch with leads
- Transparent tape
- 5 mm slow-flashing red LED
- Hookup wire
- One 68-ohm, 1/4-watt resistor
- Electrical tape
- Red embroidery thread
- Sewing thread matching the color of your stuffie
- One 3V coin-cell battery (CR2032)
- Coin-cell battery holder
- Two scraps of felt (any color): one 1" x 0.5" and the other 1" x 1.5"
- One sew-on snap

...and MAKE it!

PLACE THE SWITCH

1. With the scissors or craft knife, open a small hole in your stuffie's arm. Cut along a seam, if you can, so the arm will be easier to sew back together. Then feed the head of the tactile switch down into the arm so the leads are sticking out. (If you need to remove a bit of poly-fill to get the switch in, just stuff it back in when you're done.)

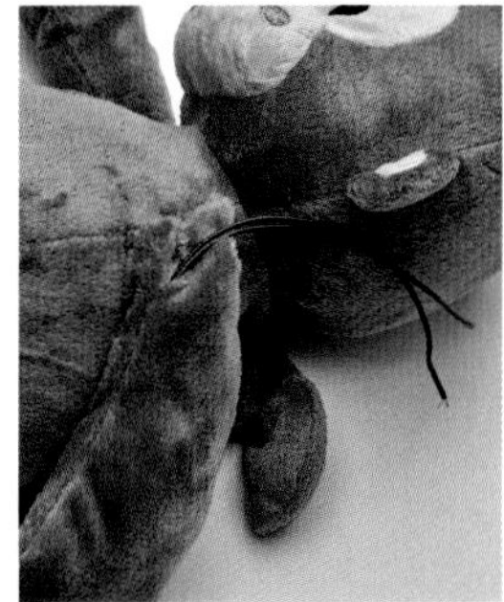

2. Carefully snip another hole in the back of your stuffie (again, on a seam, if possible). Using transparent tape, tape together the ends of the leads sticking out your stuffie's arm and thread them, like a shoelace, through the stuffie and out the back. Set your stuffie aside.

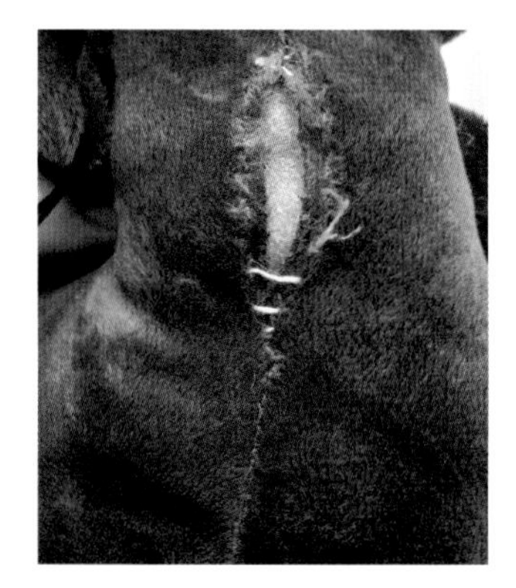

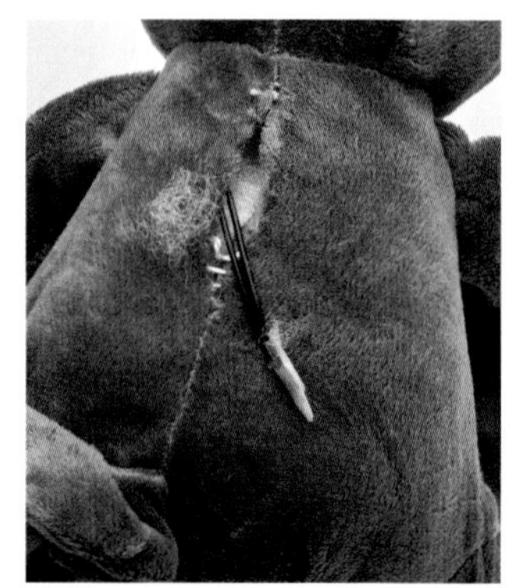

WIRE THE LED INTO THE CIRCUIT

3. Mark the long, positive (+) leg of your LED with the black marker. Then gently bend the legs of the LED apart. With the wire strippers, cut a 4" length of hookup wire and strip about half an inch off both ends. Using the soldering iron, solder one end to the negative (–) leg of your LED.

To make soldering easier, use your third-hand tool to hold the pieces in place, and hold the wires steady with your needle-nose pliers when you strip them.

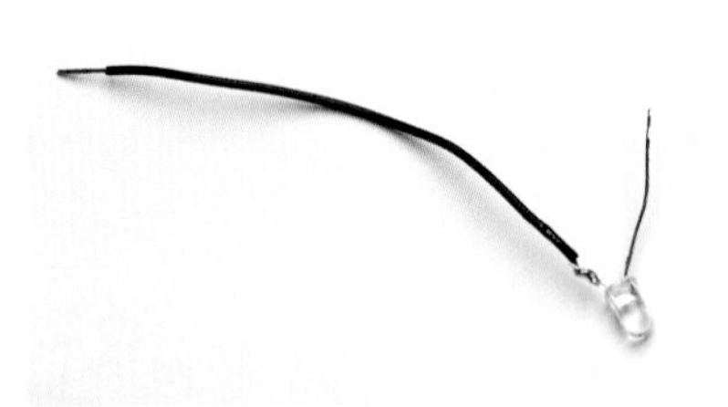

4. With your wire strippers, snip the positive leg of the LED to shorten it (this will keep the wire from flexing and breaking). Twist one end of the resistor (either end will work) onto the shortened LED leg, and solder in place. As before, use the third-hand tool to help hold the pieces.

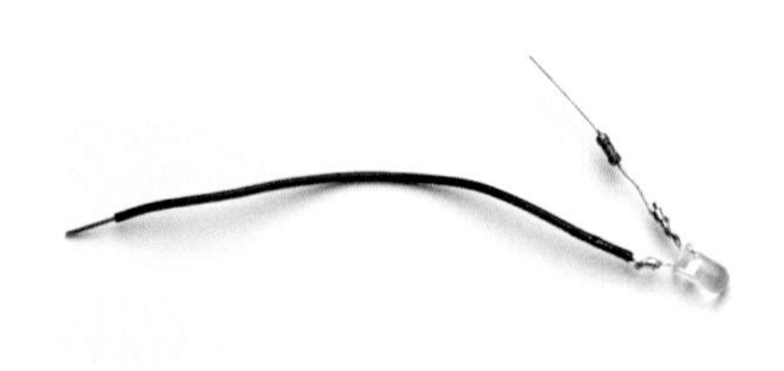

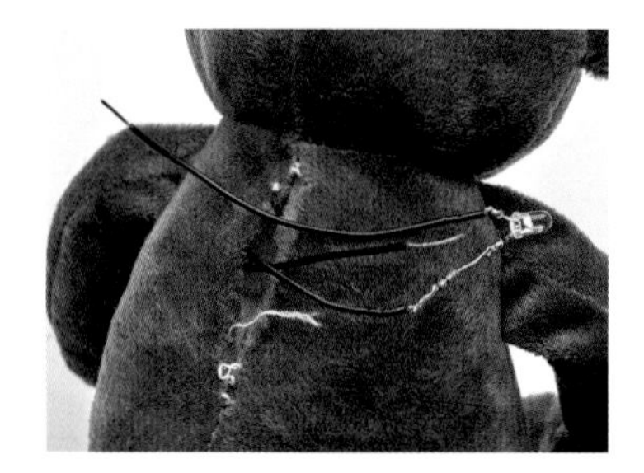

5. Using wire strippers, snip one lead of your switch (it doesn't matter which one) to shorten about an inch. Then strip that wire, twist it to the resistor's free side, and solder in place.

6. Wrap electrical tape around both sides of the LED, covering the resistor and all solder points. Using a hot-glue gun, place glue to cover the connections on both sides of the LED to prevent breaks and short circuits.

TEST IT! Hold a battery between the two wires to be sure your LED lights up.

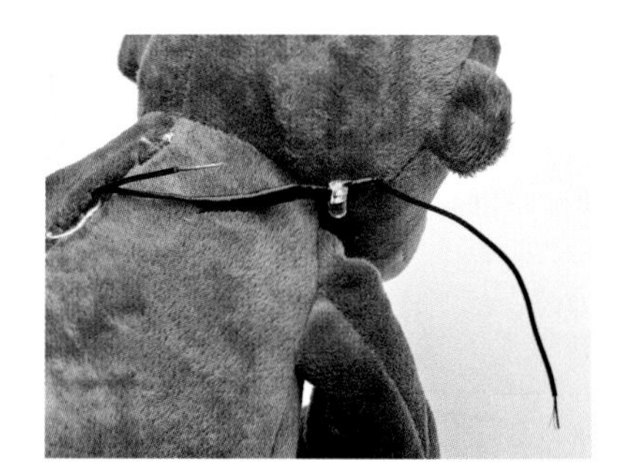

HEART YOUR STUFFIE

If you need more hands to hold the stuffie while placing the LED, ask a friend to help!

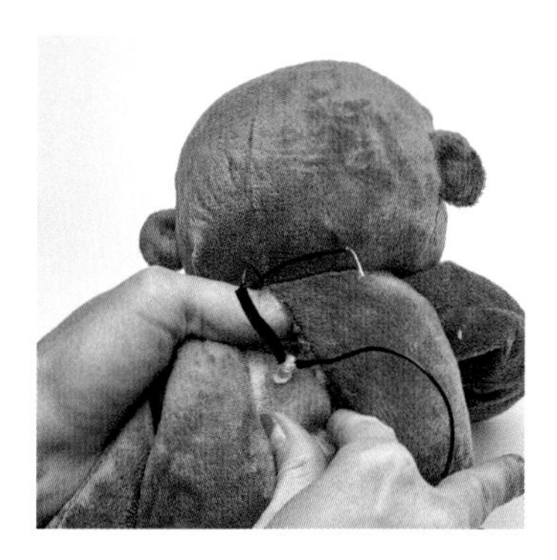

7. Wiggle your fingers inside your stuffie and figure out where its heart (your red LED) should go. Use a pencil or piece of chalk to mark the spot inside, so the mark doesn't show through to the front. Using the hot-glue gun, put a blob of glue there, wait until it's tacky, and then set the LED in place, right up against your stuffie's chest. Hold until dry.

8. If you want, as a guide while sewing, use a pencil to draw a heart shape around the LED on the inside of your stuffie. Then, thread a needle with doubled-up red embroidery thread. Start from the inside so you can hide the knots, and carefully stitch a heart around the LED on the outside of your stuffie's chest.

9. Use sewing thread and a needle to sew up your stuffie's open arm. Knot and cut the thread.

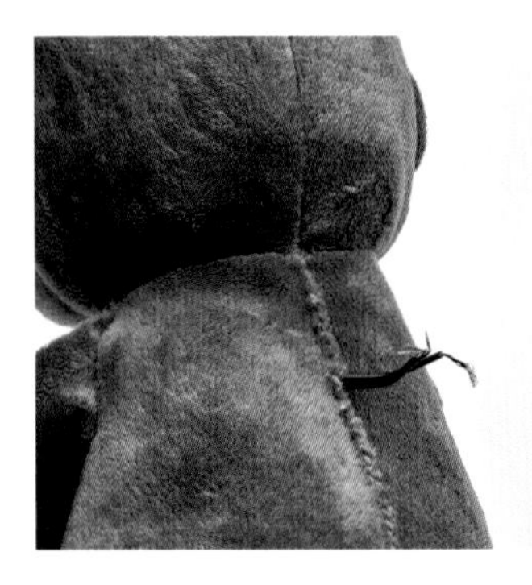

10. Use sewing thread and a needle to sew up your stuffie's back. Sew around the wires, leaving them sticking out, and leave an opening of about an inch so you'll always have access to the battery. Be sure to knot and secure the threads above and below the opening.

WIRE THE BATTERY HOLDER INTO THE CIRCUIT

11. Hold or temporarily tape the battery between the two wires. Does the LED light up on the front of your stuffie? If not, switch the battery around. When the LED lights up, note which wire is against the positive (+) side of the battery and which is against the negative (–) side. That will tell you how to orient the battery holder.

12. Twist the wires into the battery holder, top and bottom, connecting positive (+) to positive (+) and negative (-) to negative (-). (On our monkey, the top wire was the positive wire.) Slip in the battery to be sure the LED is working. When everything is oriented correctly, reinforce both joints. Using the soldering iron, solder the wires in place, and using the hot-glue gun, cover them with hot glue. Then carefully stuff the battery and battery holder into the opening.

ADD BATTERY ACCESS AND A SNAPPY FINISH

13. Using a needle and sewing thread, sew one half of the snap in the center of the small piece of felt, and the other half of the snap centered at the edge of the large piece of felt.

14. Using the hot-glue gun, glue the small piece of felt (snap side out) on one side of the 1" opening. Then, making sure the two halves of the snap align, glue the free edge of the larger piece of felt to the other side of the opening, as shown. Fold to snap the two halves together.

Press the tactile switch in your stuffie's arm to turn on its beating heart, and give it a heartfelt hug.

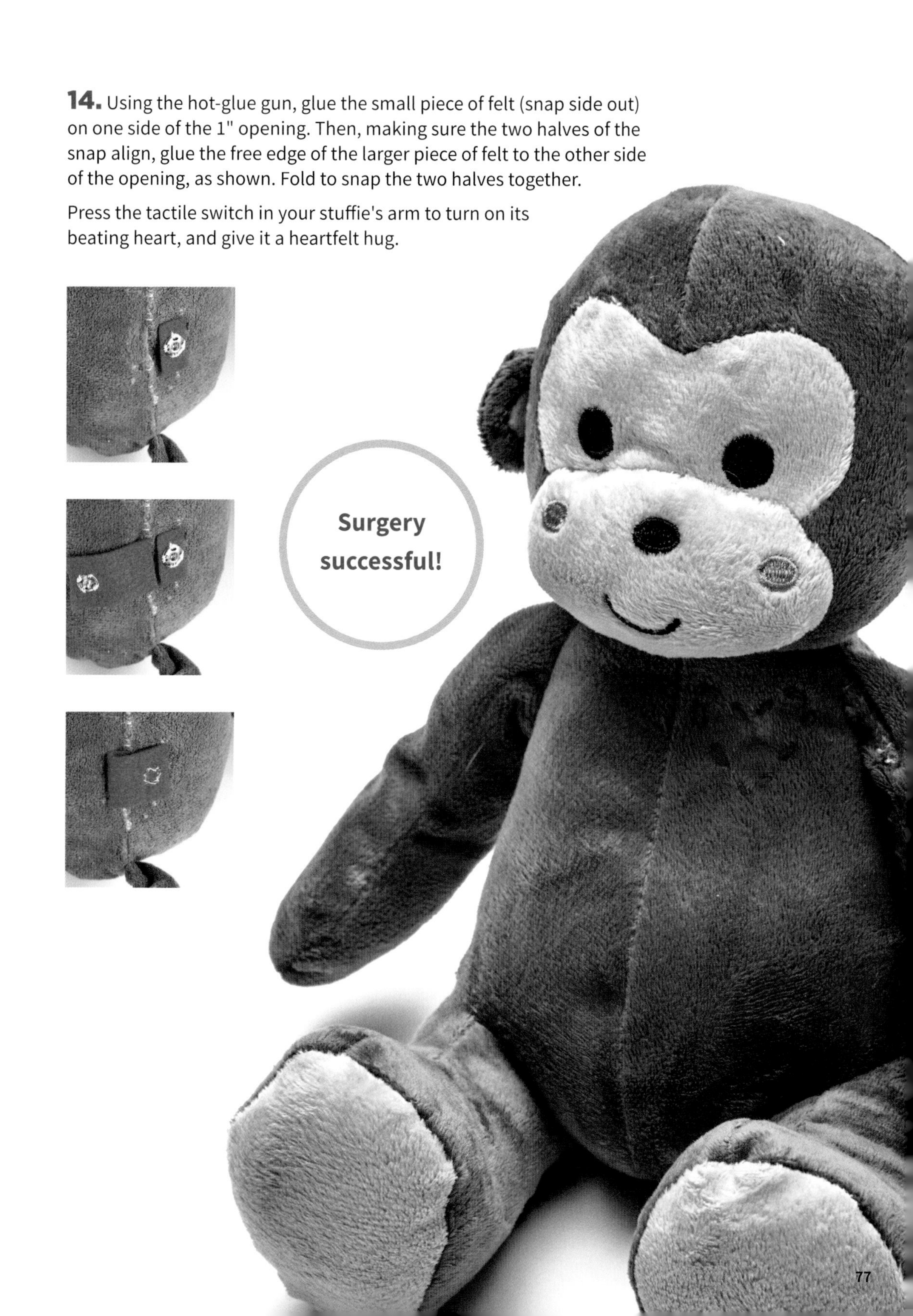

Fairy Wings

SKILL LEVEL 3

Transform a few wire hangers and some black tights into a magical set of glimmering wings.

Get your tools & materials...

TOOLS

- » Wire cutters (to cut the hangers)
- » Pliers
- » Wire strippers
- » Black marker
- » Scissors
- » Soldering iron and solder
- » Third-hand tool
- » Hot-glue gun and glue sticks

MATERIALS

- » Four bendable wire hangers
- » Black duct tape
- » One 2-cell battery holder with leads and on/off switch
- » Two pairs of adult (women's) sheer black tights
- » Hookup wire
- » Eight 5mm diffused candle flicker or flash RGB LEDs (or your choice)
- » Large paper clip
- » Electrical tape
- » Two 3V coin-cell batteries (CR2032)
- » A 3' length of black elastic, 1" wide
- » Optional: Scraps of coordinating felt

...and MAKE it!

PREP YOUR WINGS & WIRES

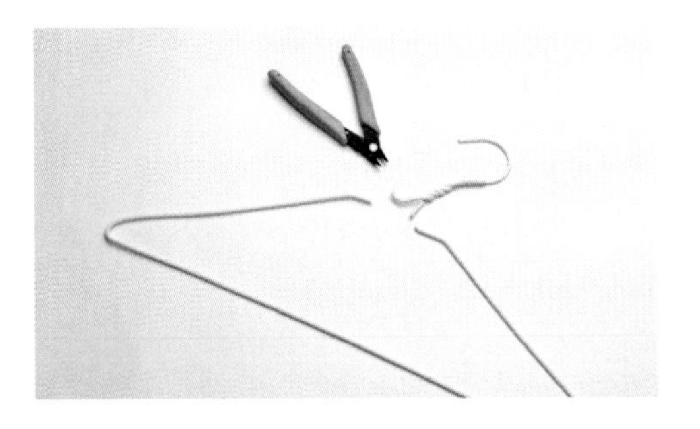

1. With the wire cutters, clip the hooks off all four hangers and discard. Straighten out each metal rod (use pliers on the corners), and then shape each of your four wing sections the way you want them to look. (Ours look a bit like squished raindrops, but yours can be rounder or flatter—whatever works best for you!)

---

2. Wrap the ends of each hanger with duct tape so they don't pop apart while you're shaping the wire wings. You should have four separate wing pieces. Set them aside.

---

3. With wire strippers, trim the leads on the battery holder to about 3" long. Then, holding the leads with pliers, strip about 1" off the ends to expose the wire inside. Set the battery holder aside.

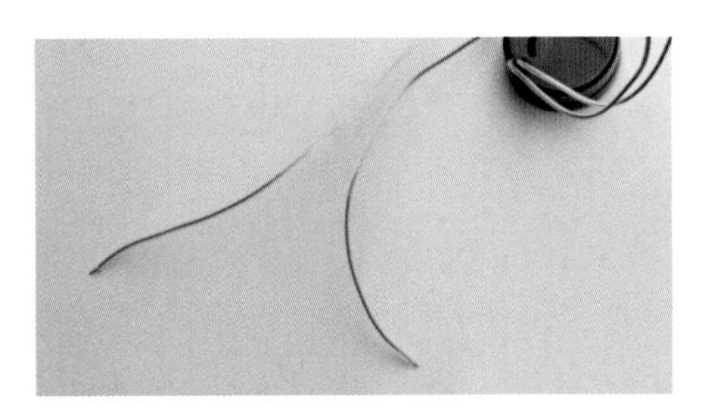

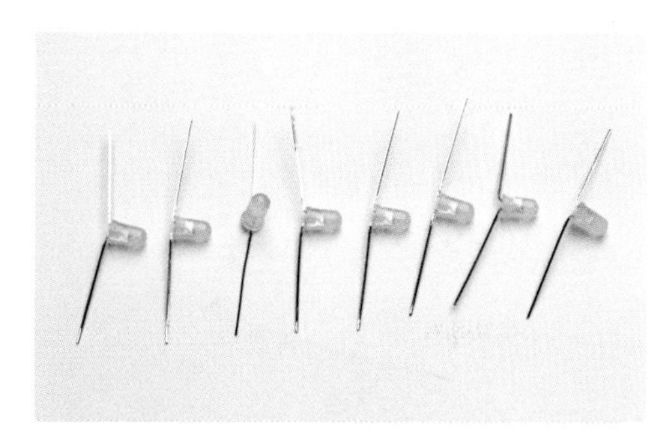

4. With black marker, darken the positive (+) legs of all eight LEDs, and then bend the legs out to the sides in opposite directions, as shown. Set aside.

5. Using scissors, cut the legs off both pairs of black tights. Discard the cut-off briefs (or save for another project), and put the four loose legs aside.

WIRE-UP YOUR TOP WINGS

Note that we're wiring LEDs into the two top wings only.

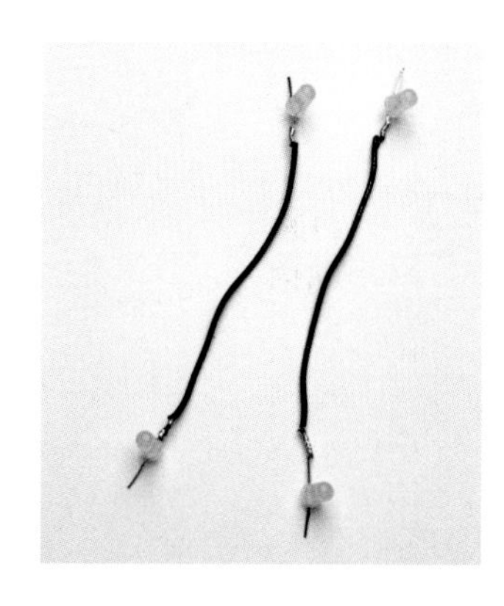

6. Using wire strippers, cut two pieces of hookup wire, about 12" each, and strip the ends to expose about 1" of wire. These will be the main positive (+) and negative (–) leads for one of your wing shapes. Take the first piece of wire and, using the soldering iron, solder the negative (–) leg of one LED to one end. Then solder the positive (+) leg of another LED to the other end. Do the same thing with the second piece of hookup wire.

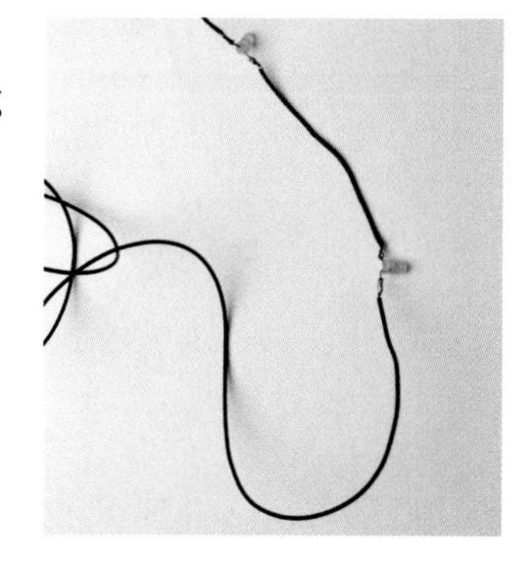

7. Using wire strippers, cut four lengths of hookup wire about 18" to 24" long and strip the ends. Using the soldering iron, solder one of these to each of the four unsoldered LED legs on the wire-and-LED assemblies you created in step 6. Use the third-hand tool to help hold the wire while you solder. This will make two assemblies, each with two LEDs and three lengths of wire, as shown. On each, tape a big paper clip to the positive (+) end, so you can keep track of the polarity.

8. Using one of the top wing shapes, wrap one of the long wire assemblies around it, so that the two LEDs appear in the middle of the shape, positioned as shown on page 82. Using electrical tape, tape the wires to the metal hanger to hold them in place as you work.

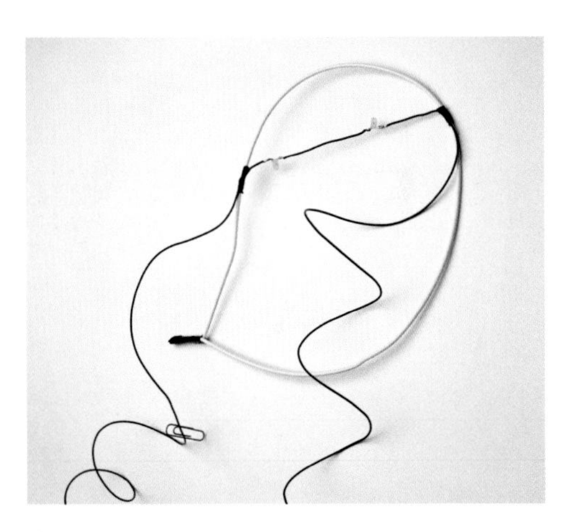

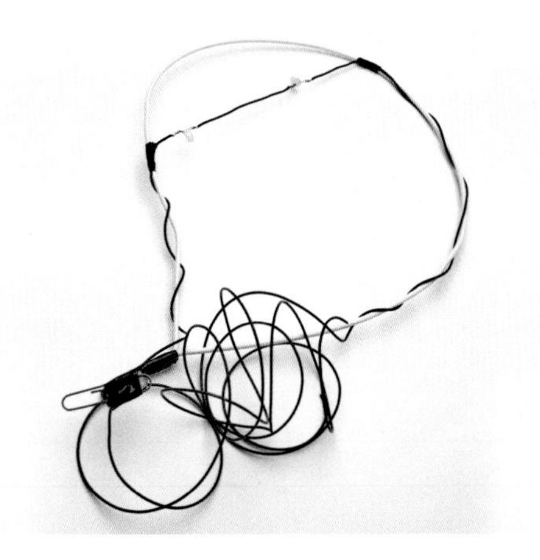

9. Add a second long wire to the wing you've been working on (the top wing from step 8). Position the LEDs so that they appear at different places within the wing shape, wrapping and taping as you go. Make sure both positive (+) ends of each wire wrap around the same side of the wing, and the negative (–) ends wrap around the other side. Using a taped-on paper clip, hold the two positive (+) wires together, and set this wing section aside.

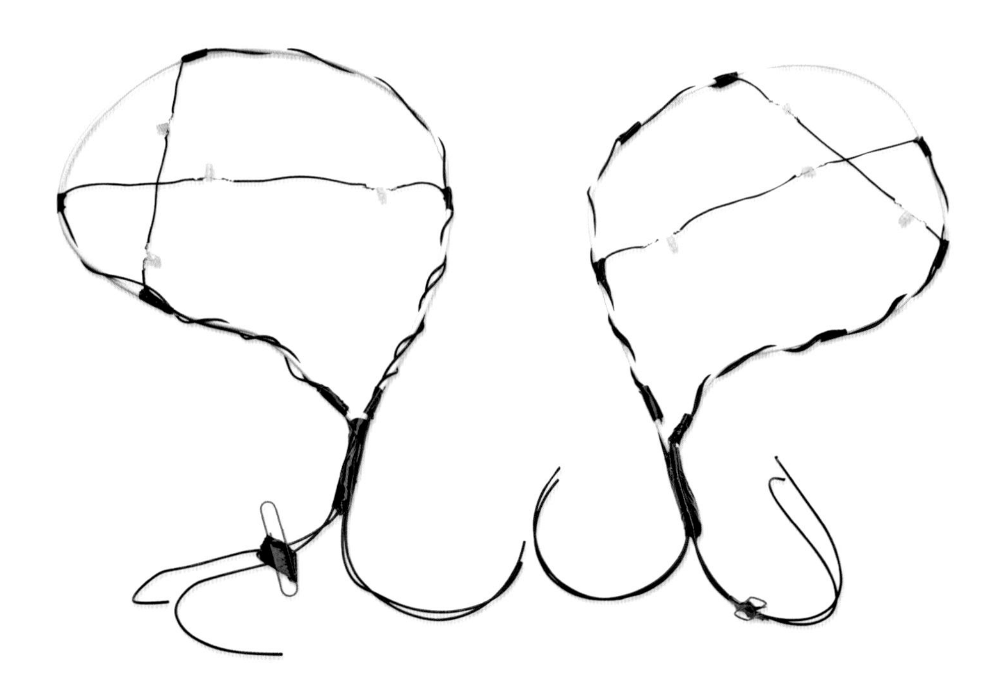

10. To assemble the second top wing, repeat steps 6 through 9.

LIGHT THEM UP

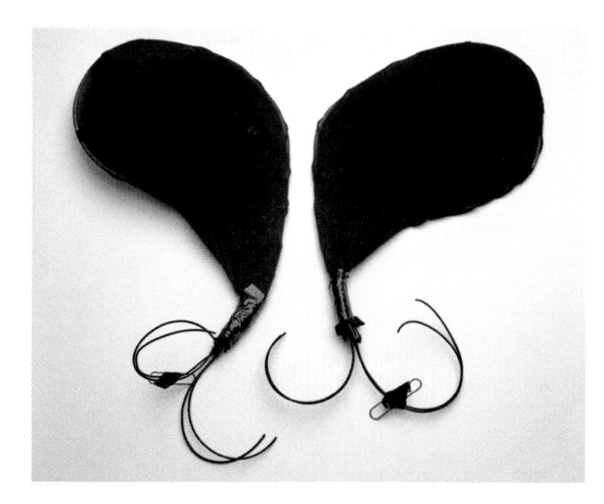

11. Take one of the cut-off legs from the black tights and pull it gently over one of the top wing assemblies. (Be careful—the LEDs can snag and rip the material!) Knot, trim, and tape the tights in place with black duct tape at the bottom of the wing, as shown. Then do the same with the other top wing assembly, covering it with another leg from the cut-off tights.

12. When both top wings are covered in the tights, use black duct tape to tape both top wing segments together, as shown. If necessary, reshape the hanger wire so the wings look the way you want them to.

13. Gather the four positive wires on one side of your assembly (the ones marked with paper clips) and twist the exposed wire ends together. Then do the same thing with the four negative wires.

TEST IT! Load both batteries into the battery holder and touch the leads to the two twisted-together wires. All your LEDs should light up.

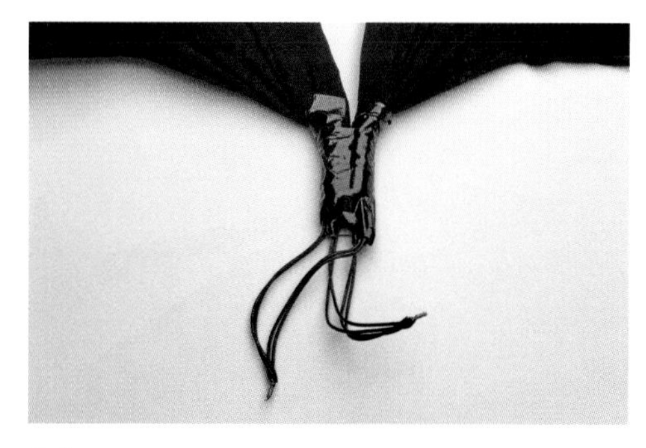

14. When everything is working, use the soldering iron to solder each twist of exposed wire together. Work from one side, then flip and work from the other side to be sure each thick bundle of wire is secure.

SWITCH THEM ON

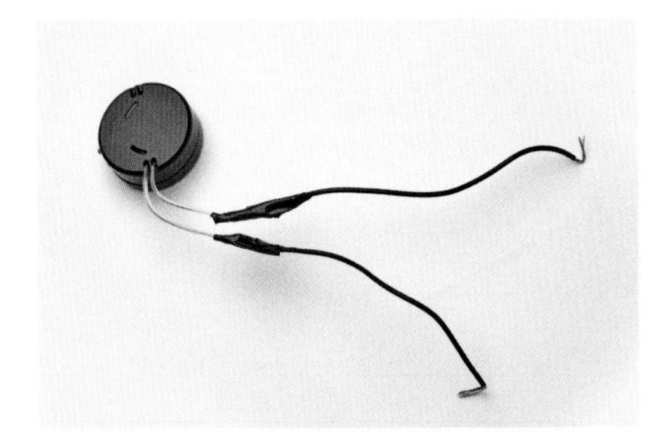

15. Using wire strippers, cut two 4" pieces of hookup wire, strip the ends, and then solder one wire to each of the two leads on the battery holder.

16. When everything is working, solder the battery holder's leads to the wing assembly's wiring by connecting the positive (+) lead to the positive (+) wires and the negative (-) lead to the negative (-) wires.

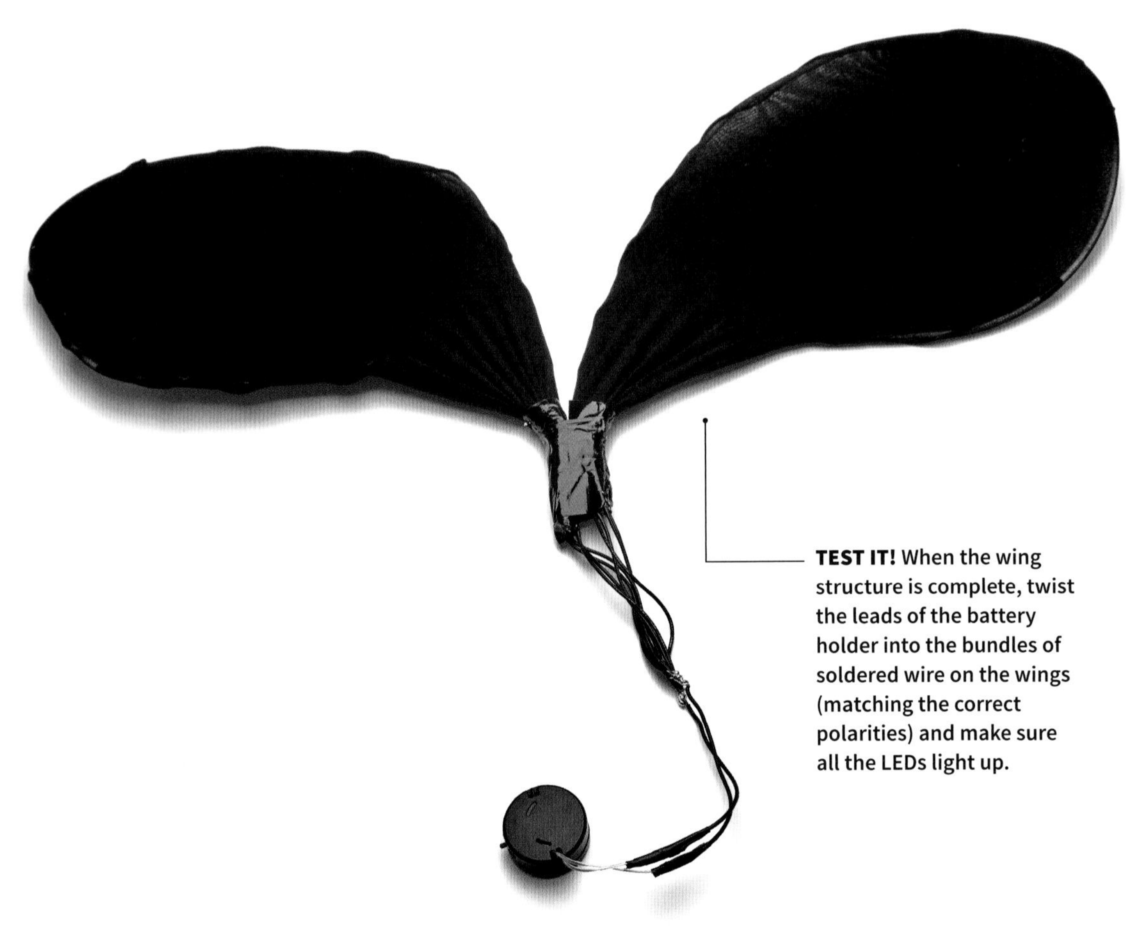

TEST IT! When the wing structure is complete, twist the leads of the battery holder into the bundles of soldered wire on the wings (matching the correct polarities) and make sure all the LEDs light up.

PUT IT ALL TOGETHER AND TAKE WING!

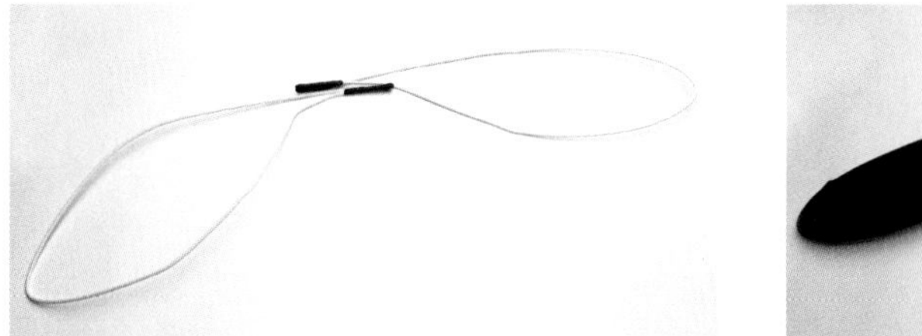

17. Make the bottom wings by pulling the last two cut legs from the black tights over the two remaining wire hanger assemblies. Knot, trim, and tape the tights to the ends of the wing frames, as shown, and then shape these new wing sections however you like them.

---

18. Using black duct tape, tape the bottom and top wing assemblies together. When everything is in place, gather up the wires and tape in a neat bundle in the center of the wings. If you want to, you can cover the battery assembly with a scrap of felt, using either glue or black duct tape.

---

19. Make two shoulder straps by looping lengths of elastic around your shoulders. Using scissors, cut the elastic where the fit is comfortable. Using the hot-glue gun, glue the ends of each length together. Then use hot glue to connect the two loops in the middle.

20. When your shoulder straps are done, tape them to the center of the wings, adjust fit, and switch everything on.

You're on your way to Fairyland!

Superhero Cape

SKILL LEVEL 2

Not even Superman had a cape this cool. For this project, a friend can be helpful to finalize the design.

Get your tools & materials...

TOOLS

- » Pencil
- » Scissors
- » Chalk or other fabric marker
- » Hot-glue gun and glue sticks
- » Sewing needle
- » Black marker

MATERIALS

- » Cardstock for design template (old file folders work great)
- » One or two pieces of colored felt (light colors work best)
- » Plain-colored T-shirt (an adult size Large is usually best)
- » Regular thread
- » Sew-on Velcro
- » Up to five sew-on sequin LEDs
- » Large sew-on snap
- » Sew-on battery holder
- » Conductive thread
- » Embroidery thread
- » Clear nail polish

...and MAKE it!

CREATE YOUR SUPERHERO EMBLEM

1. Start by deciding on a superhero emblem. We made a big M (for Maker), but you can use anything you want! Try out your own initial, choose a number (is there a special birthday coming up?), or create your own design (a lightning bolt? a star? a circle with an LED face?).

Whatever you choose, use a pencil to draw the emblem on your cardstock, and then use scissors to cut it out to make a template for your design.

2. Using chalk or a fabric marker, trace your emblem onto a light-colored piece of felt and use scissors to cut it out. You don't have to use two colors, as we did. A simple emblem cut from one piece of felt will work (light gray works well, since it matches the conductive thread). If you do decide to use two layers, use the hot-glue gun to glue them together. Decide where you want to put the LEDs within the emblem, so you'll be ready to place them when you start sewing.

GET THE T-SHIRT READY

3. Fold the T-shirt in half lengthwise (sleeve to sleeve) and lay it flat on a table. The cape is made out of the back of the shirt—but don't remove the T-shirt's neck!

With chalk, draw one half of your cape's outline along the T-shirt's folded side. With scissors, cut from the bottom of the T-shirt, following the chalk line, cutting through both sides of the shirt. Cut away the front of the shirt by trimming around the neck ribbing, keeping it intact.

4. Next, find the center of the neck ribbing and cut through it, as shown. Put the cape on to test how it hangs and how big you want the neck to be. If the neck is too big, use scissors to snip off as much extra ribbing as necessary. Then, using a needle and regular thread, sew tabs of Velcro onto each end of the ribbing, so you can easily put the cape on and take it off.

STICK, SNAP, AND SEW

5. With chalk, mark small dots on the back of your felt emblem where the LEDs will go. (If it's hard to see through the felt, just hold it up to the light.)

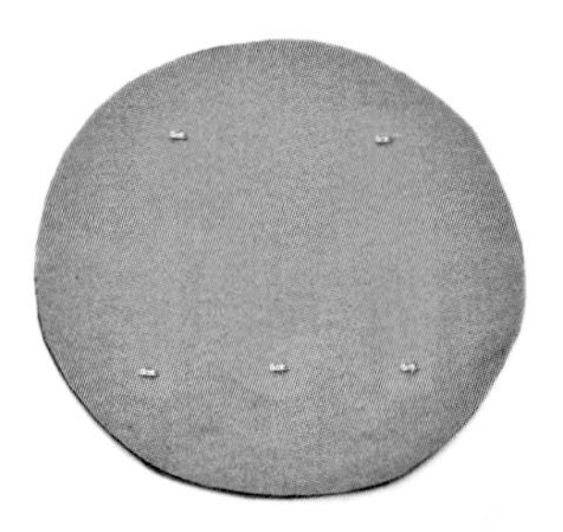

With a black marker, darken the positive (+) side of each LED on the back, so you can identify them. Then, using dots of glue, stick the LEDs in place. Put them bulb-side-down (so they'll shine through the fabric), with the bulb in the glue. Make sure, for each LED, that the negative (–) side is to the left and the positive (+) side is to the right.

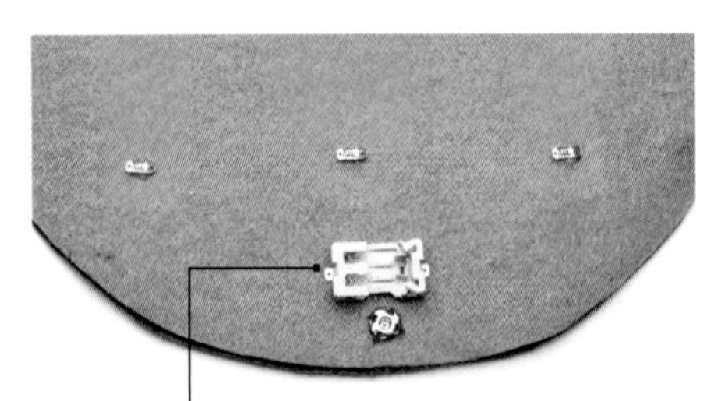

6. Using a needle and regular thread, sew the female side of the snap on the back of your felt emblem. Position this about half an inch from the bottom in the middle. Then use the hot-glue gun to glue the battery holder in place just above the snap.

We've attached the battery holder with the positive (+) side to the left and the negative (–) side to the right. You can position it either way as long as you don't let the conductive threads touch or cross over each other, shorting out the circuit.

7. Using a needle and conductive thread, sew the female side of the snap to the positive (+) side of the battery holder (as shown, this is on the left of our holder, which may be hard to see). The snap will be your switch, so it must be connected to the rest of the circuit. Then, without cutting the thread, continue sewing from the holder's positive (+) side through the positive (+) sides of all the LEDs in your design. When you're done connecting all the LEDs, knot, cut, and trim your thread.

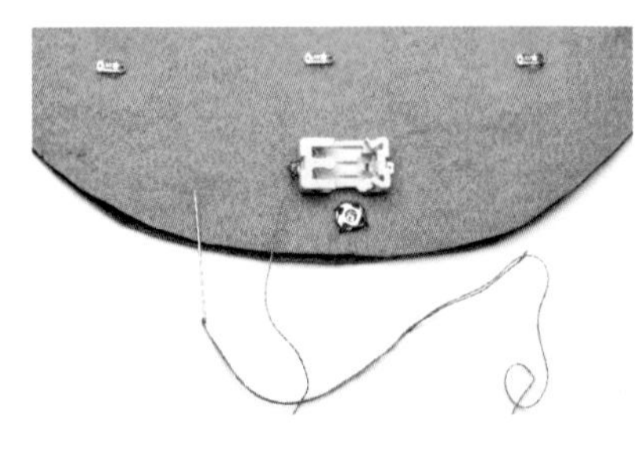

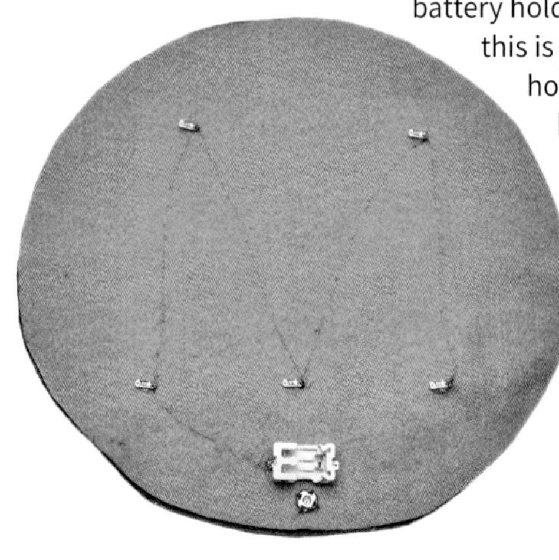

8. Using a needle and conductive thread, sew to connect the negative (–) sides of all your LEDs. Begin with the first LED, not the battery holder, and sew until you get to the last LED. Knot in place, but this time leave a long length of uncut thread dangling free; remove the needle. In step 12, you will use this thread to connect to the male side of the snap.

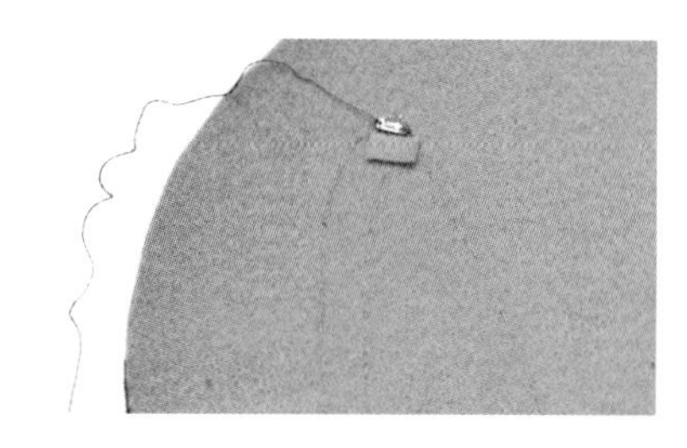

As you sew the negative (–) lead, if you get to a place where you need to cross over the positive (+) line of sewing (as we did), use the hot-glue gun to glue down a scrap of felt as a "bridge," just like in the Power Cuff project (page 58). Then sew across the bridge (not all the way through it!) to keep the threads separate.

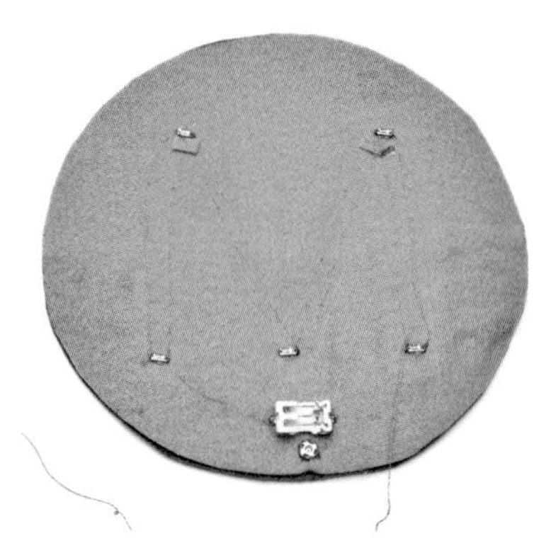

PUT IT ALL TOGETHER

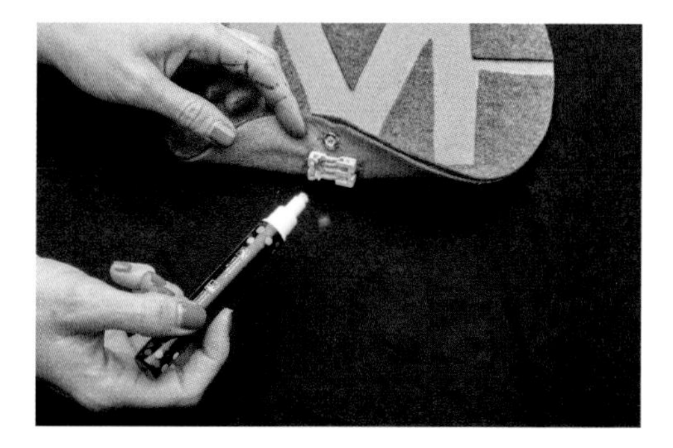

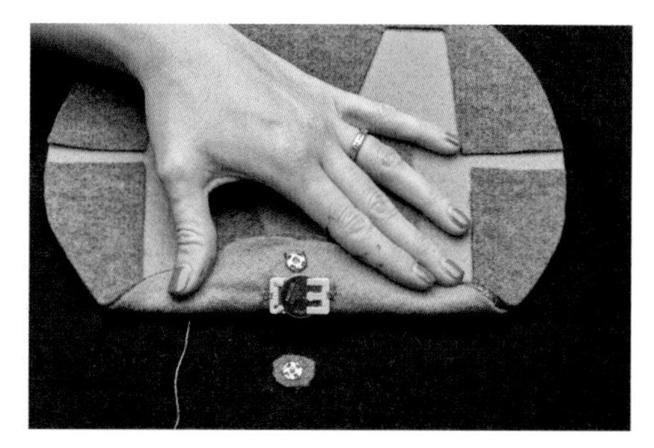

9. To decide on the best place for your emblem, ask a friend to help you model the cape. Once you figure out the exact position for your emblem, note where the male side of the snap should be to meet the female side at the bottom of the emblem. With chalk, mark that spot on the cape.

10. Using a needle and regular thread, sew the male side of the snap to a small scrap of felt. Then use the hot-glue gun to glue the snap onto your cape at the chalk mark.

11. If you used a double layer of felt to make your emblem, you may want to use a needle and embroidery thread to stitch it together now, so that it will look nice when you're done. Otherwise, use a hot-glue gun to glue your finished emblem in place on the cape, leaving the bottom few inches unglued so you can get to the battery pack. Make sure the loose end of conductive thread is not glued down and is easy to reach.

12. Rethread a needle with the loose end of conductive thread (which is from the negative side of the LEDs). Stitch the thread through the cape itself (that is, the T-shirt material) until you sew the thread into the male side of the snap.

By sewing the snap into the circuit, you're creating a simple switch. When you snap on the emblem, you'll complete the circuit and the LEDs will turn on!

TEST IT! Slip in the battery and make sure everything is working. If so, cover all your conductive-thread sewing with clear nail polish to protect it.

That's it! When you're ready, snap the snap, put on your cape, and test out your new superpowers!

Light-up Tote Bag

SKILL LEVEL 3

If you've been wondering just what all that stuff is at the bottom of your tote bag, this project will solve the problem. (It also comes in handy on Halloween.)

Get your tools & materials...

TOOLS

- Craft knife
- Third-hand tool
- Soldering iron and solder
- Ruler or measuring tape
- Wire strippers
- Screwdriver (for the reed switch)
- Scissors
- Hot-glue gun and glue sticks
- Chalk
- Seam ripper
- Sewing needle

MATERIALS

- One cut-to-length 8" LED light strip
- Canvas tote bag
- Red hookup wire
- Black hookup wire
- Two-piece magnetic reed switch with "Normally Closed" option
- One 9V battery snap
- Scrap felt or fabric
- One 9V battery
- One ¾" magnetic purse clasp
- Regular thread to match tote bag
- Optional: light-colored fabric or felt

...and MAKE it!

WIRE THE LED STRIP

1. On the 8" LED strip, find the two tiny negative (–) and positive (+) copper soldering pads at one end. Use a craft knife to gently scrape off the surface of both. Then, using the third-hand tool to hold the LED strip, prepare the two copper pads by using the soldering iron to drop a bit of solder on each. (This technique is called "tinning": Having a bit of metal on both sides of a connection creates a stronger bond and makes soldering easier).

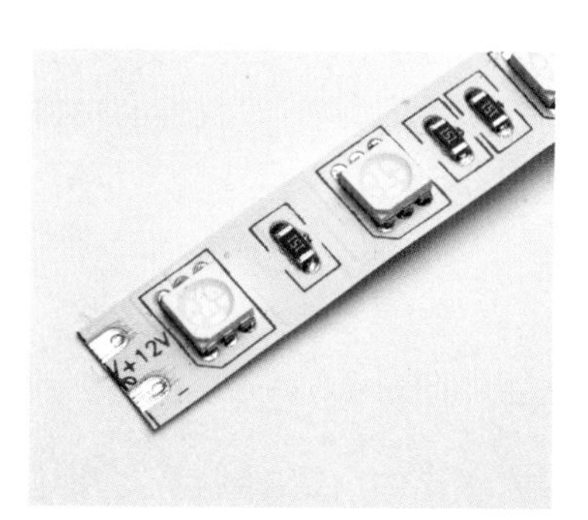

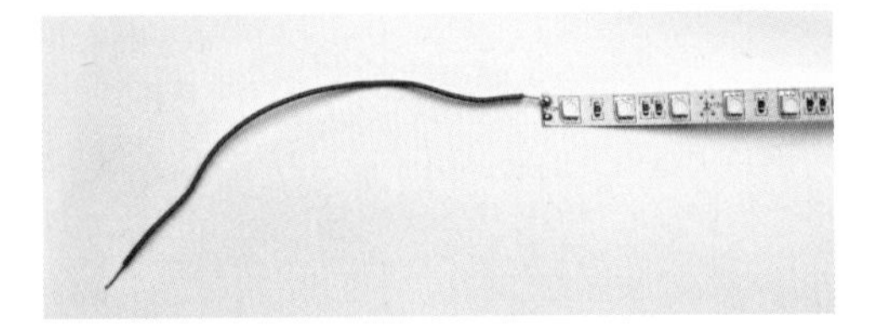

2. Place the LED strip on the inside hem of your tote bag, centered between the handles on one side.

Using a ruler or measuring tape, measure from the tinned end of the LEDs to the side hem of the tote and add 1" to the measurement. Use wire strippers to cut a piece of red hookup wire to that length. Strip both ends, exposing about an inch of wire and, using a soldering iron, solder one end to the tinned positive (+) pad of your LED strip.

---

3. Reposition the LED strip as before, centered between the tote bag handles, and measure from the same tinned end of your LED, past the side seam, and around the hem until you get to the center point between the handles on the opposite side of the bag. This is where the reed switch will go. Add 1" to that measurement and, using wire strippers, cut a piece of black hookup wire to that length. Strip both ends and solder one to the tinned negative (–) pad of your LED strip.

Your LED strip should now have one short red wire and one long black wire attached, like this:

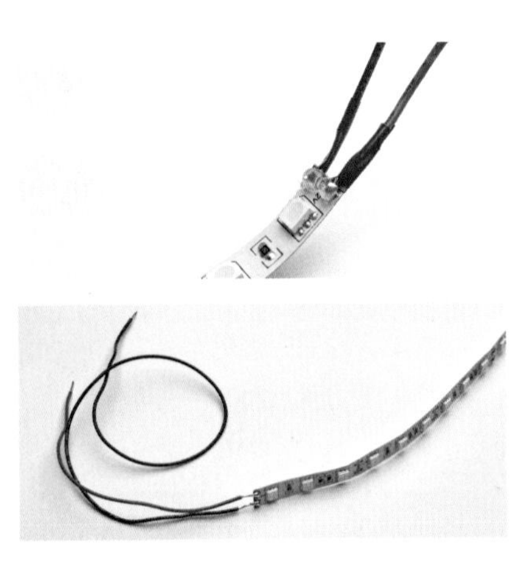

ADD THE REED SWITCH AND BATTERY CONNECTIONS

4. Separate the two halves of the reed switch. Set the unmarked half aside (you'll use it in step 15). On the other half, find the NC ("normally closed") setting and loosen that screw with a screwdriver. (On our switch, it's the one in the middle, as shown.) Wind the loose end of the black hookup wire around the screw, and tighten the screw again.

In its "normally closed" setting (used here), the switch stays in the open ("off") position as long as the magnet is against the switch. That means when you open your tote bag, the switch changes to its closed position, allowing power to flow from the battery so the lights can come on.

5. Using wire strippers, cut a piece of black hookup wire about 6" long and strip both ends. On the reed switch, unscrew the COM ("common," or ground) setting, wind one end of the 6" black hookup wire around the screw, and tighten the screw back up.

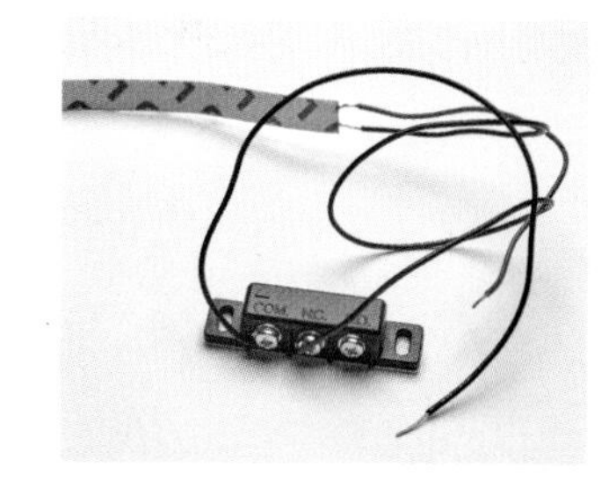

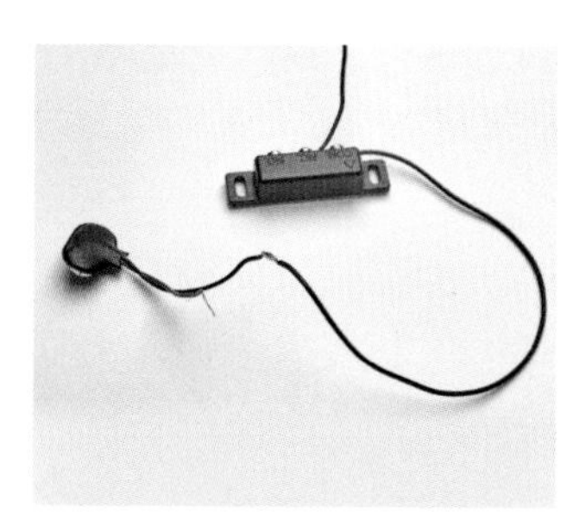

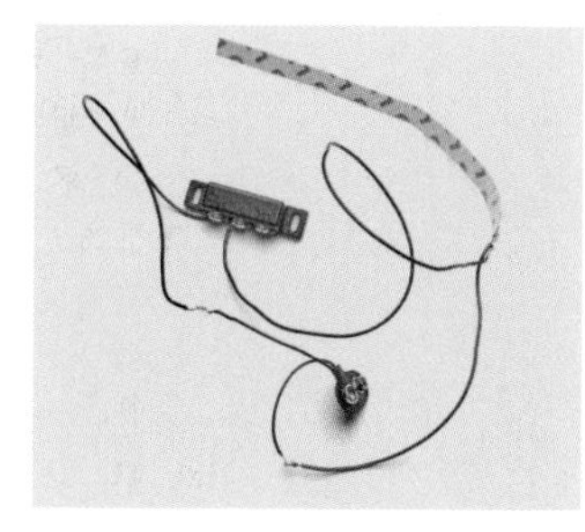

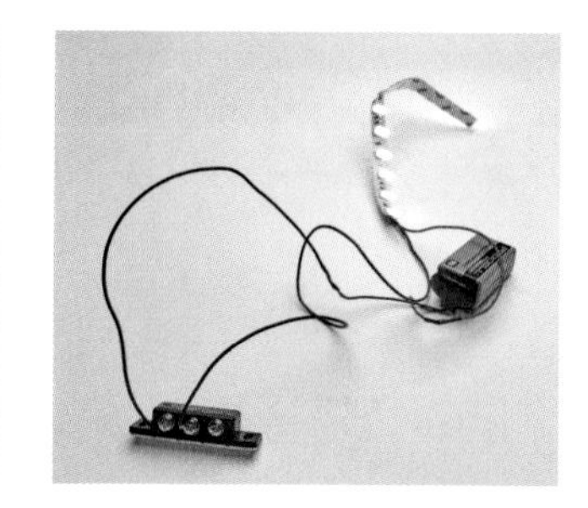

6. Take the 9-volt battery snap (which will snap onto the top of the 9-volt battery) and, using wire strippers, strip the ends of its wire leads. Using the soldering iron, solder the battery snap's black wire to the free end of the 6" black wire attached to the COM setting on the reed switch (in step 5).

Once the circuit is working, remove the battery, and set this assembly aside.

TEST IT! Plug the 9-volt battery into the battery snap and your LEDs should light up!

MAKE A SEW-IN BATTERY POCKET AND PREP YOUR PURSE SNAP

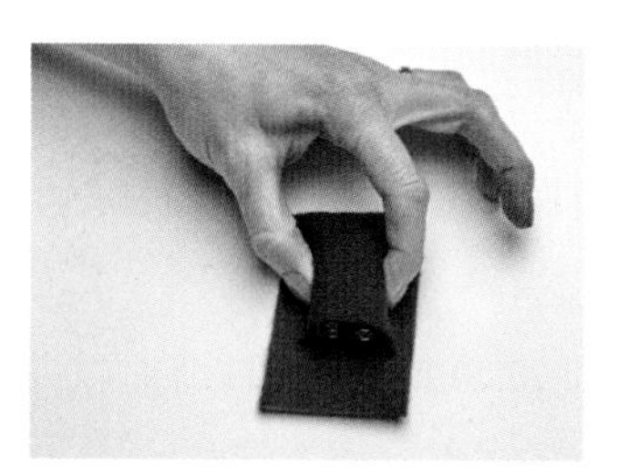

7. To make the battery pocket, using scissors, cut two pieces of felt: one should be about 5" x 2.5" and the other about 2" x 2.25". Lay the 9-volt battery lengthwise on top of the larger piece and bend the smaller piece over the battery. Using a hot-glue gun, place glue around three sides of the larger felt piece and press them against the smaller pieces, tucking in the edges. Leave the top open with a lip of felt, as shown. Set aside.

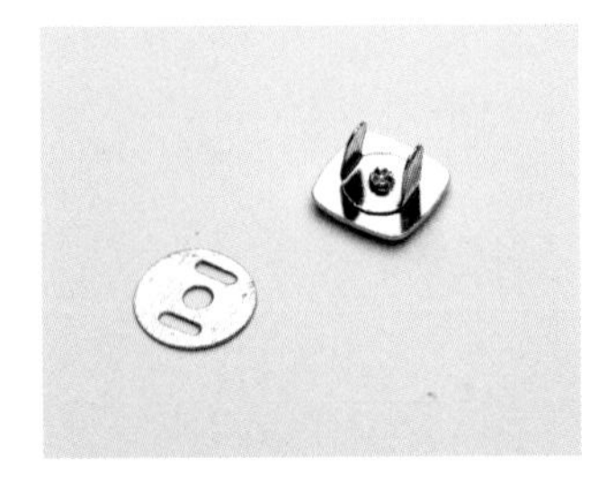

8. Using scissors, cut a tab of felt ¾" wide and as tall as the hem on the tote bag. Snip two slits in the felt and insert the legs of one side of the magnetic purse clasp. Add the washer over the felt and bend the legs of the clasp down. Repeat with the other half of the clasp. Set both aside.

ADD THE CIRCUIT TO THE TOTE

9. With chalk, mark the center point between the handles on both sides of the tote. These marks will help you center the LED strip on one side, and they mark the placements for the snaps of the magnetic purse clasp on either side of the tote.

10. Using the seam ripper, rip open the inside hem of your tote bag. Start at the end of the LED strip with the wired assembly and continue around the tote bag to the opposite side, ending at the center point between the two handles. This is where you will hide the wires.

11. On the side of the tote where the LEDs will go, center the LED strip (using the chalk mark on the tote). Then, using a hot-glue gun, glue the strip in place.

12. Feed the wires in under the open hem, flattening the wires as much as possible as you go.

Using the hot-glue gun, glue to close the seam and keep the wires in place—but make sure to leave the 9-volt battery snap hanging out in the middle! Stop gluing when you get to the chalk mark on the opposite side of the bag, leaving the switch sticking out.

13. Next, glue one side of the magnetic purse clasp (it doesn't matter which one) on top of the LED strip. Center the clasp between the tote's handles at the chalk mark. Your bag should now look like this.

14. On the opposite side of the bag from the LED strip, center the reed switch at the chalk mark, place it just below the tote's hem, and use the hot-glue gun to glue it in place.

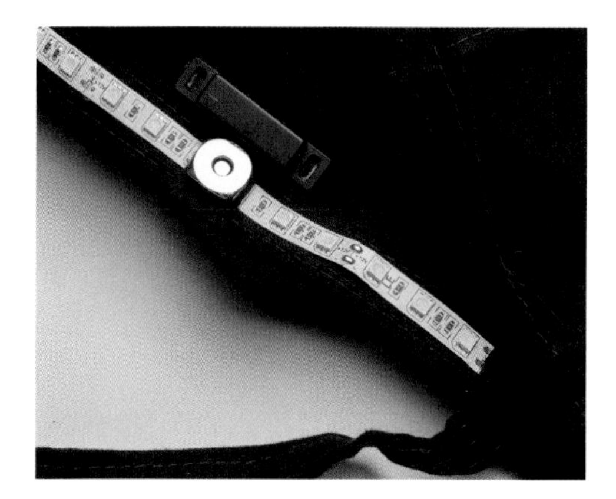

15. Take the unmarked half of the reed switch (which you set aside in step 4), and align the arrows on both halves so they are facing in the same direction. Once aligned, use a hot-glue gun to glue the unmarked half in place just below LED strip and the purse clasp.

Then take the other half of the magnetic clasp and glue it in place above the marked half of the reed switch (the one with the attached wires).

16. Finally, using a needle and regular thread, sew the battery pocket to the inside of the tote bag wherever it fits best.

When everything is in place and working correctly, you can stitch the hem back together again. Add the battery and check out your tote. If the lights are too bright, just glue a strip of light-colored fabric or felt on top to diffuse it.

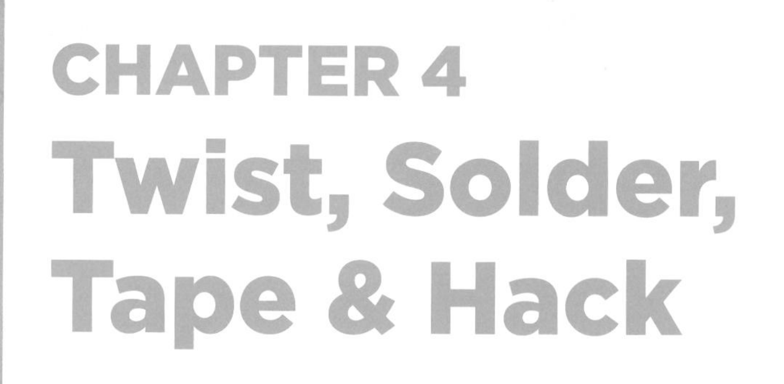

CHAPTER 4
Twist, Solder, Tape & Hack

This chapter mixes things up with activities that range from simple to challenging, but all offer unexpected ways to use tools and materials. Electric Roses (page 104) mixes paper and wires, Thunderclouds (page 107) repurposes holiday lights to dramatic effect, and the Glue Jewel projects (beginning on page 120) use glowing globs of glue to jazz up jewelry.

Now look around and imagine what you can do to glow up your world.

Electric Roses

SKILL LEVEL 1

There's nothing more charming than a lovely bouquet of bright-colored blooms. But even if you don't have a green thumb, you can make this memorable "glowquet" (glowing bouquet) to brighten up any occasion. This project lists the materials for one electric rose, so just multiply to make as many as you want!

Get your tools & materials...

TOOLS

- Wire strippers
- Scissors

MATERIALS

- One 10" length of red hookup wire
- One 10" length of black hookup wire
- One gumdrop LED
- Electrical tape
- One premade fabric or paper flower
- One 3V coin-cell battery (CR2032)

...and MAKE it!

MAKE THE STEM

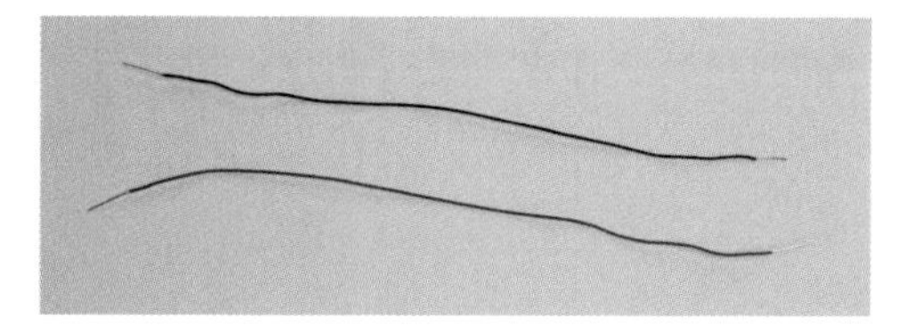

1. Using wire strippers, strip about an inch off both ends of the red and black wires.

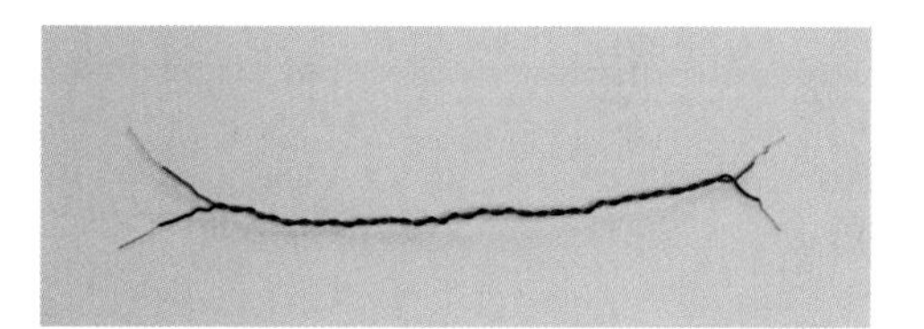

2. Twist the red and black wires together to make the flower stem. When you're done, separate the ends of the wires at each end.

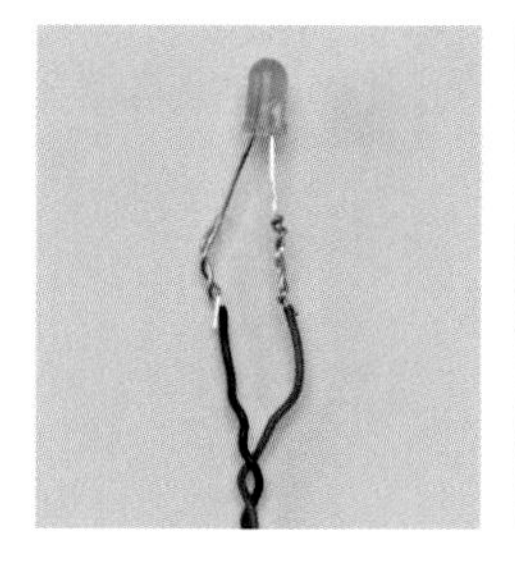

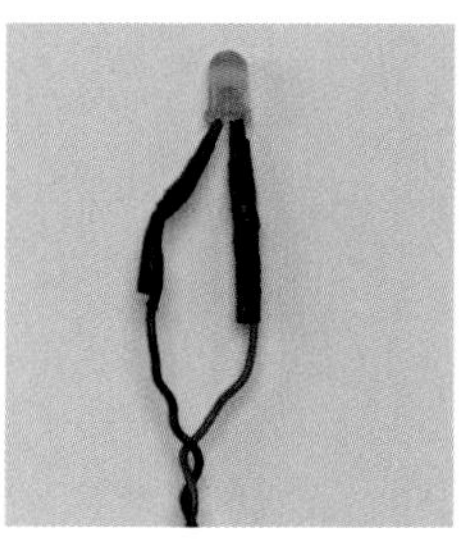

3. Carefully attach the gumdrop LED so that its long positive (+) leg is attached to the red wire and its short negative (–) leg is attached to the black wire. Wrap each connection separately in electrical tape.

Be sure the positive (+) and negative (–) wires don't touch, or you can short out the circuit!

ASSEMBLE THE FLOWER

4. Gently pull the wire stem down through the top of the premade flower so that the LED nestles into the center of the flower.

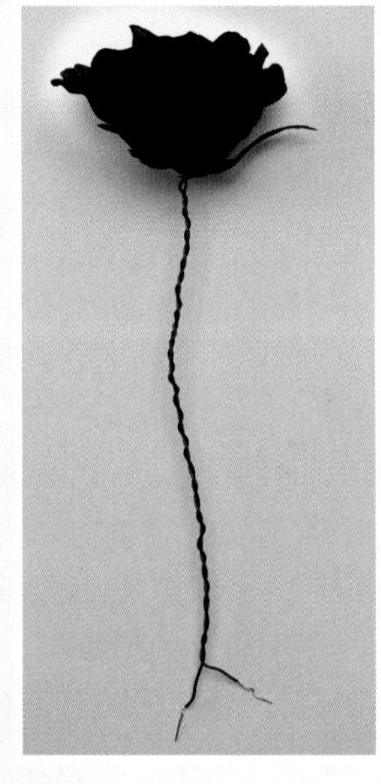

5. Position the battery between the two stripped wires at the bottom of the stem. Place the red positive (+) wire against the smooth, positive (+) side of the battery and the black negative (–) wire against the bumpy, negative (–) side, and wrap the battery in electrical tape to hold everything place. The LED in the center will light up.

Someone sweet will love these! Make eleven more for an even dozen.

Thunderclouds

SKILL LEVEL 1

Love a good storm? Make your own! These DIY Thunderclouds offer all the drama—but none of the dangers—of one of nature's most dramatic displays. Depending on the type of LEDs you choose, you can make them gleam, twinkle, or burst with flashes of LED lightning. Easy to make and fun to personalize, they'll transform any scene from serene to spellbinding.

Get your tools & materials...

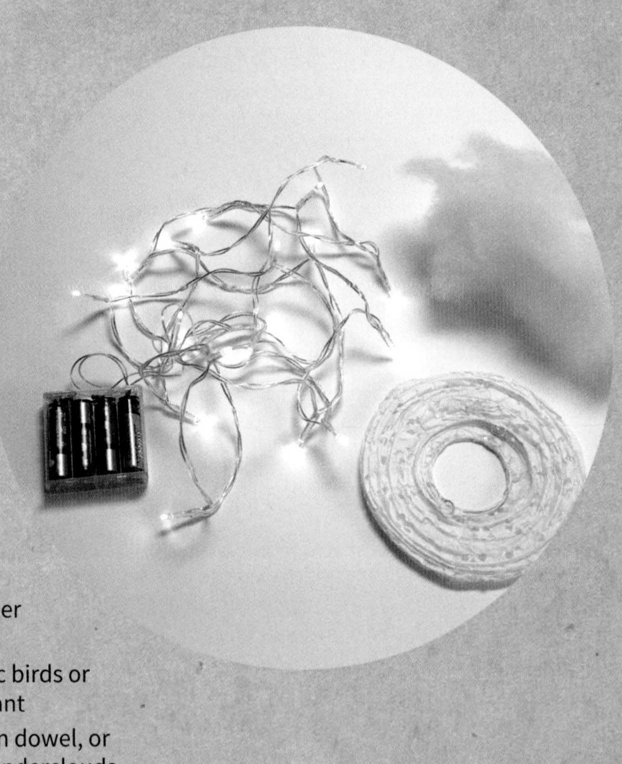

TOOLS

- Hot-glue gun and glue sticks
- Scissors

MATERIALS

- One paper lantern (you can buy them singly or in packs of different sizes)
- Bag of fluffy white polyester fiber filling
- An LED source: string of white LED holiday lights preprogrammed to blink/flash/stay steady (either plug-in or with battery pack and batteries), Glowies, or other
- Transparent packing tape
- Fishing line, picture-hanging wire, or other sturdy, invisible thread
- Optional: glitter, spray paint, little plastic birds or airplanes—whatever decorations you want
- Optional: screw-in ceiling hooks, wooden dowel, or PVC pipe for hanging bigger, heavier Thunderclouds

...and MAKE it!

FLUFF UP YOUR LANTERNS...

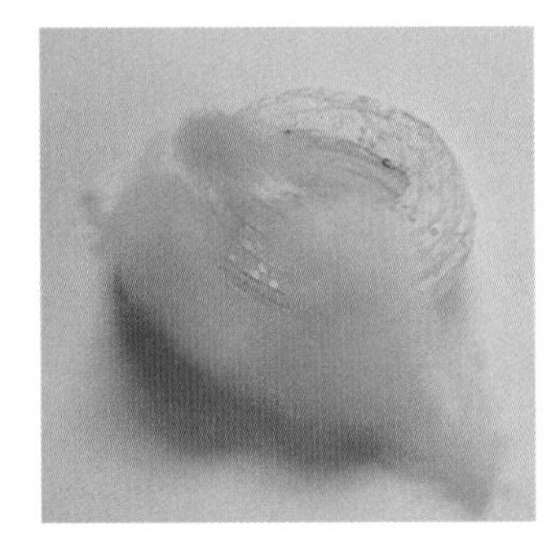

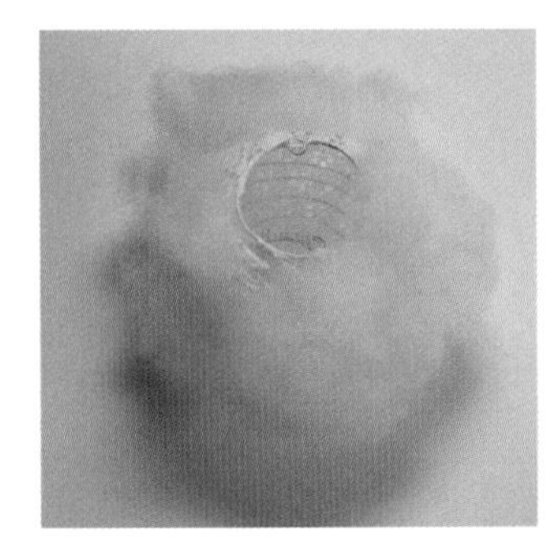

1. Open up the paper lantern (they usually come flattened out). Using a hot-glue gun, glue puffs of polyester fiber filling all over the outside, so the lantern is covered. Apply the glue in small sections, and smush on the fiber so your lantern starts to look like a cloud.

...AND ADD THE LIGHTNING

2. When the whole lantern is covered with polyester filling, put in the Glowies or arrange an LED light string inside the lantern. Use tape to secure the end of the light string to the rim of the lantern, making sure you can easily reach the battery pack (if there is one). Depending on the LED source you're using, you make be able to tape the battery pack inside, too, to keep it hidden from sight.

---

3. Flip on the LED lights and see how it looks! If you want to decorate further, now's the time. Pull a few strands of lights through the bottom of the lantern to make flickering rain. Add a splash of glitter to twinkle things up, or hit it with a few bursts of spray paint to highlight storm-darkened edges. Use your imagination!

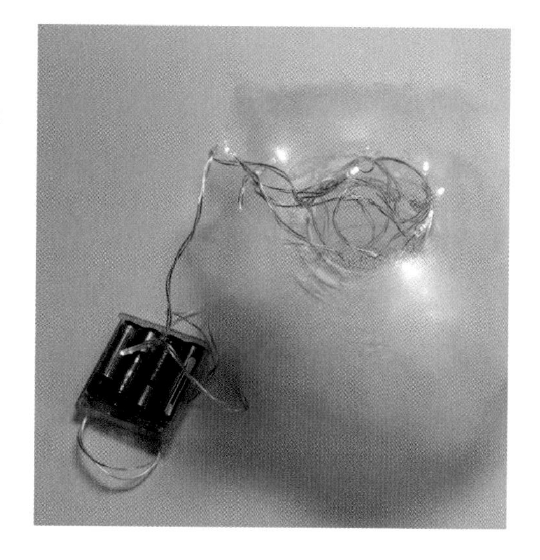

---

4. When you're done, have a grownup help you hang your cloud. We used translucent fishing line, but picture wire or string could also work.

If you've made a big cloud from lots of stuck-together lanterns, or want to hang many different clouds, one way is to attach them (with grownup help) to a wooden dowel or length of PVC pipe using line or wire. First hang the dowel or PVC pipe securely from the ceiling using ceiling hooks, and then tie the clouds to the dowel or pipe. Just be sure you don't bore holes into the living room ceiling without adult permission—and help.

Firefly Nightlight

SKILL LEVEL 1

If you'd rather play with bulbs than bugs, these bottled-up "fireflies" are great! They'll glow for hours on end, creating a colorful nightlight that will spare the innocent fireflies of the world, while lighting up the night with a twinkly glow.

Get your tools & materials...

TOOLS

- Black marker
- Scissors
- Optional: needle-nose pliers

MATERIALS

- Conductive thread
- Several 3V coin-cell batteries (CR2032)
- Electrical tape
- Several blinking LEDs (and a few nonblinking if you want)
- See-through jar with a screw-on lid

...and MAKE it!

BUILD A BUG...

1. With the marker, darken the long, positive (+) legs of all your LEDs so you can easily keep track of them.

2. Using scissors, cut two lengths of conductive thread a bit shorter than the inside of the jar. Use electrical tape to attach one piece of thread to the smooth, positive (+) side of the battery, and the other piece of thread to the bumpy, negative (–) side of the battery.

3. Choose one LED and tape its long, positive (+) leg to the thread attached to the battery's positive (+) side. Then tape the LED's shorter, negative (–) leg to the thread attached to the battery's negative (–) side.

Now that you've made a complete circuit, your LED should light up: You've made your first firefly.

THEN MAKE MORE AND BOTTLE THEM UP

4. Tape the battery with your lit-up LED to the inside of the jar's lid.

---

5. Add another firefly to your assembly by twisting the legs of a second LED around the conductive thread. You can use your fingers to bend the legs into place, or use needle-nose pliers, if you prefer. Be sure you attach the LED's long, positive (+) leg to the positive (+) thread and the LED's short negative (–) leg to the negative (–) thread, as before. Secure with electrical tape.

---

6. Use more batteries and LEDs to make as many fireflies as you want, taping each new battery to the top of the jar, and twisting in extra LEDs where you can. Mix up blinking and nonblinking LEDs for a realistic look, but be careful not to crowd the LEDs. Try to place them along the edges of the jar so the conductive threads don't cross one another and short out your circuits. (If you want, fold strips of electrical tape lengthwise over any exposed conductive thread.)

When you're done, screw on the lid, and watch your electronic bugs twinkle and glow through the night.

Marquee Letter

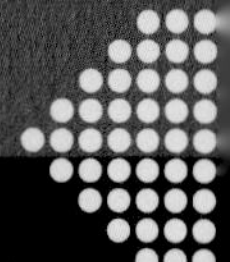

SKILL LEVEL 3

Like an old-time movie marquee, these luminous letters add spark to any space. Personalize your room with a number, a letter, even a short message spelled out in lights.

Get your tools & materials...

TOOLS

- Craft knife
- Ruler
- Pencil
- Black marker
- Wire strippers
- Hot-glue gun and glue sticks
- Soldering iron and solder
- Paintbrush
- Electrical tape
- Optional: sandpaper
- Optional: additional decorative elements (stickers, cut-out images, glitter, paint, and so on)

MATERIALS

- Hollow cardboard letter (or papier mâché letters) from craft store
- Acrylic spray paint, any colors
- Diffused LEDs, enough for size of letter (see step 2)
- 150-Ohm, ¼-watt resistors (same amount as LEDs)
- Red hookup wire
- Black hookup wire
- One two-battery coin-cell holder with leads
- Two 3V coin-cell batteries (CR2032)
- Optional: double-sided foam tape

...and MAKE it!

PREP YOUR LETTER

Note that hollow letters often have a core of extra cardboard inside; just throw this away.

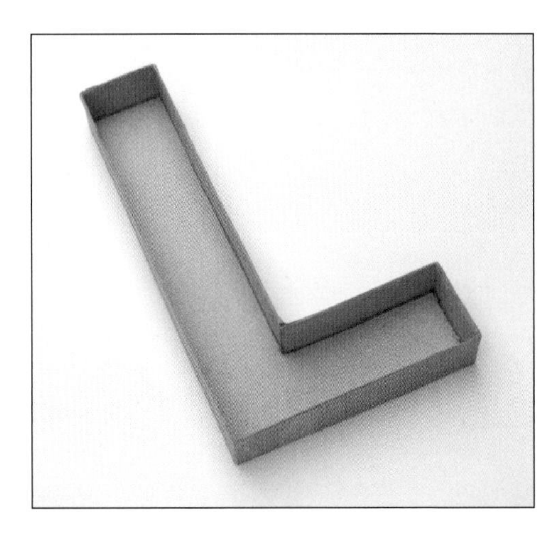

1. With a craft knife, cut off the top of the cardboard letter. It's easiest to make a rough cut first, and then follow up with a second cut to get closer to the edges. Once you've finished cutting, you can smooth the edges of the letter with sandpaper, if you want.

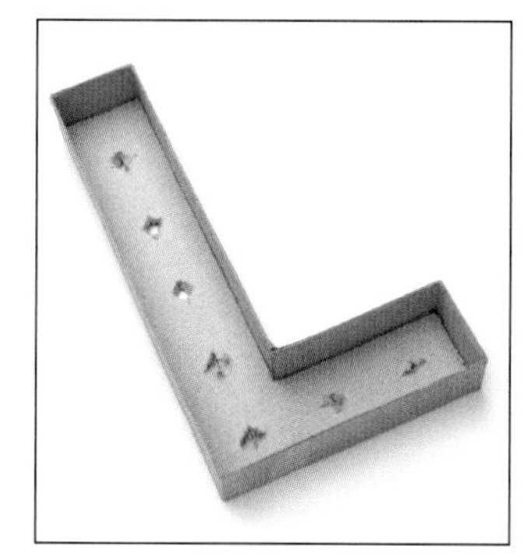

2. Using a ruler, on the back of your letter, measure where you want the LEDs to go. They should be about 1.5" apart. With a pencil, mark a small X at each spot.

3. Using the craft knife, carefully cut a small crossed slit at each X. Then, gently poke the pencil through each slit, going from inside the letter to the back, to make a rounded hole for each LED.

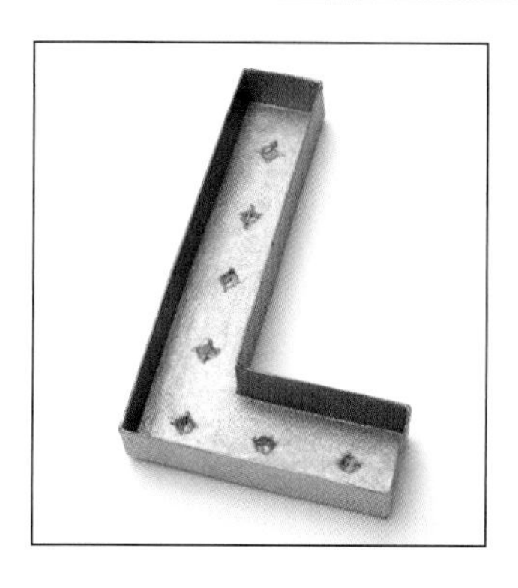

4. When the holes are made, it's time to decorate! The hollow inside is what people will notice. First paint it with acrylic spray paint, using the paintbrush to touch up the cardboard edges or add decorative details. Then add stickers or cut-out images for decoupage, sprinkle it with glitter—whatever you want. We painted ours silver to give it an old-timey look, then we painted the outside black. When you're done, let everything dry before continuing.

ADD YOUR LEDS AND RESISTORS

5. Use a black marker to darken the long positive (+) legs of each LED, so you can easily see them. Then use the wire strippers to clip each darkened leg so it's only about ½" long.

Then use the wire strippers to clip one leg of each resistor (it doesn't matter which one), so that each is about ½" long. Set the resistors aside.

6. From the front, gently wiggle one LED into each hole. Then, on the back, bend down each LED's short negative (–) leg and orient them so they're facing in the same direction. Using the hot-glue gun, glue each LED in place. Glue from the letter's (hollow) front, while pressing the X cuts in place on the back; this is to hold the LEDs tight and to keep the glue from running down onto the LED legs.

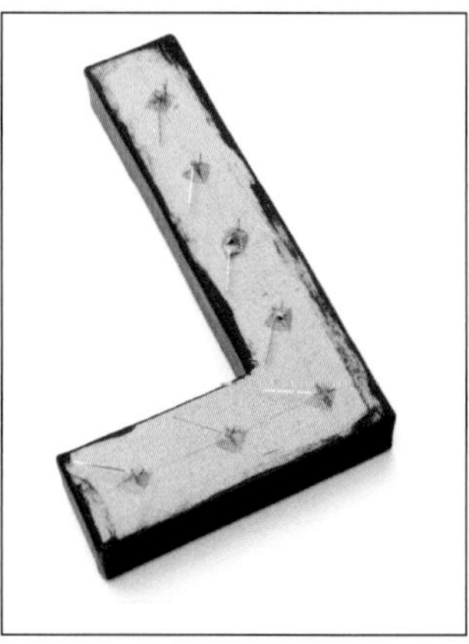

When you're done, each LED bulb should be sitting flat inside the front hollow of the letter, and each negative (–) LED leg should be pressed flat and aligned on the back.

7. When the glue is dry, flip over the letter. On the back, twist the clipped end of each resistor to the clipped positive (+) leg of each LED.

Using the soldering iron, solder each connection. Bend the wires down flat and, using wire strippers, cut off any wire that sticks out beyond the outer edge of the letter.

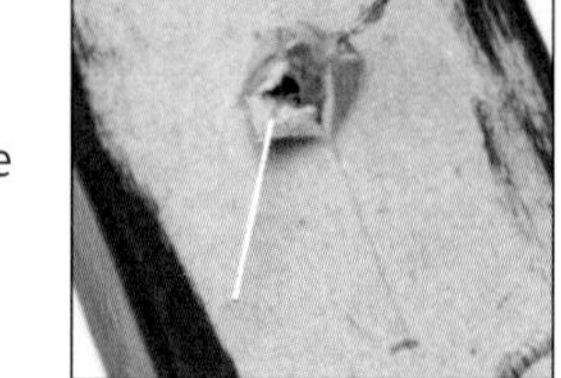

We're using resistors here because we're wiring in parallel, using more battery power than any one LED can handle. The resistors will "resist" the extra voltage so the LEDs don't burn out.

WIRE IT UP!

8. Using wire strippers, cut a piece of red hookup wire that is long enough to connect all the LEDs, plus a few additional inches. An exact measurement isn't necessary; you just need some extra wire at the end to attach the battery holder. Strip both ends to expose about 1" of wire.

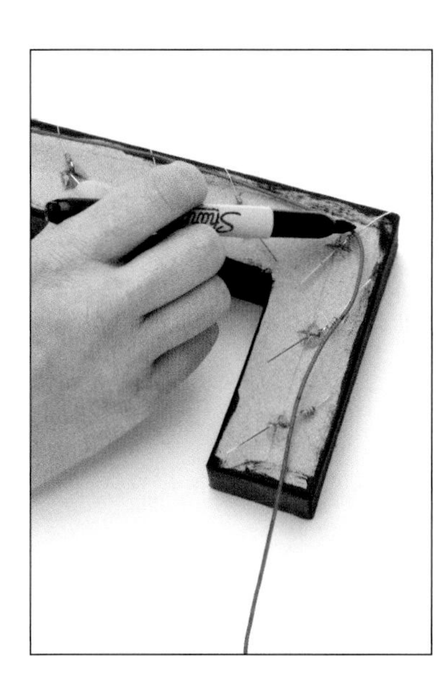

9. Hold the red wire along the LEDs and, with the black marker, mark each place the wire meets an LED. At each of these points, you want to strip enough covering away to expose a small bit of wire, which will be solder points. A quick way to strip the insulation in these places is to gently roll the wire under the blade of the craft knife, then slice lightly along the length of the wire and pull off the insulation. When finished, the red wire should look something like this:

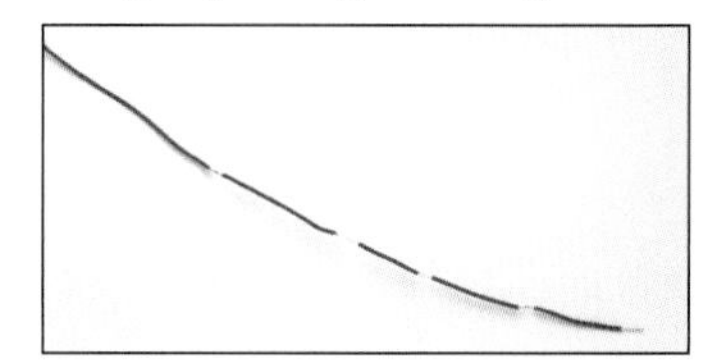

If you accidentally cut through the wire, don't worry about it. Strip the insulation from each end, and work with the same red wire, just in smaller pieces. The solder will hold everything in place.

---

10. Repeat steps 8 and 9 with the black wire, so that it measures and looks the same as the red wire, except that the black wire will connect to the negative (–) legs of your LEDs.

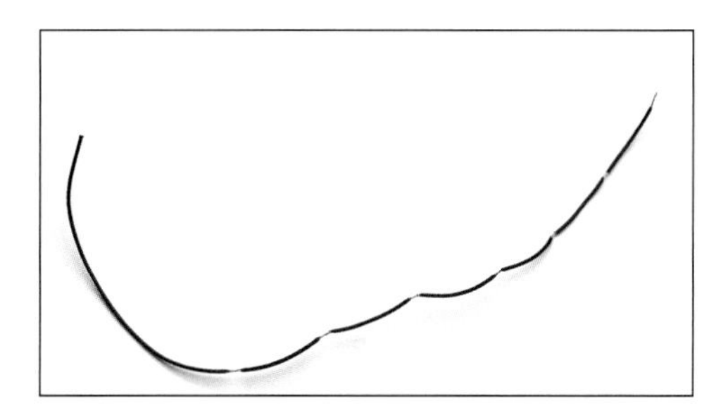

---

11. When both wires are ready, use the soldering iron to solder the red wire to the free ends of each resistor. Then solder the black wire to the negative (–) LED legs.

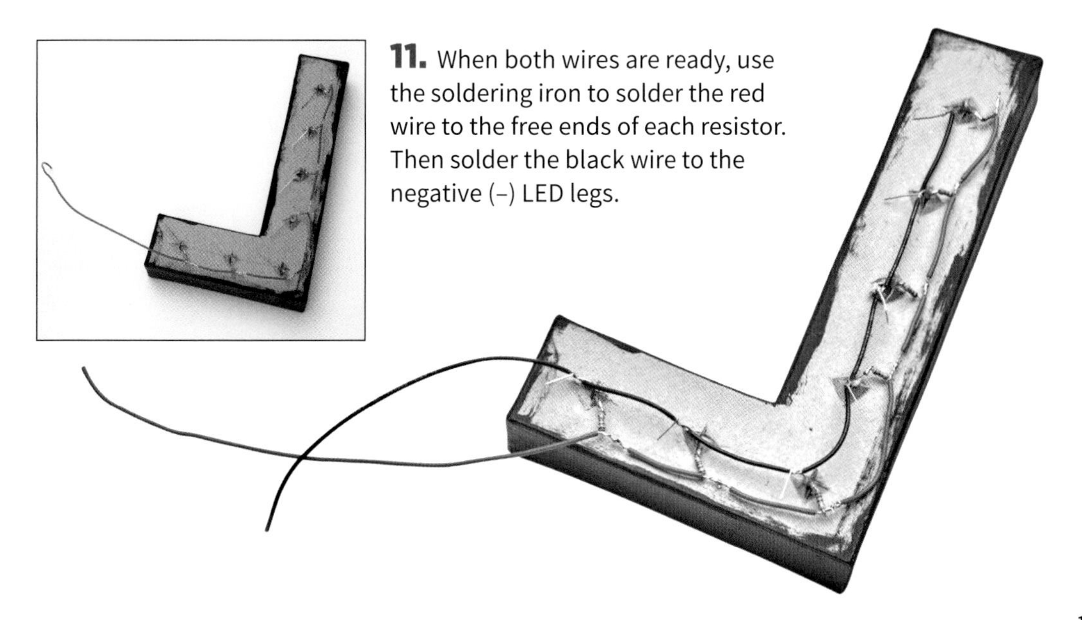

ADD THE BATTERY HOLDER AND TIDY UP

12. Using the wire strippers, clip off any extra red or black wire at the end of your letter. If the leads on the battery holder aren't already stripped (some come that way), use the wire strippers to strip the ends of the red and black leads on the battery holder. Then use the soldering iron to solder the battery holder leads to the letter's two wires: match red to red and black to black.

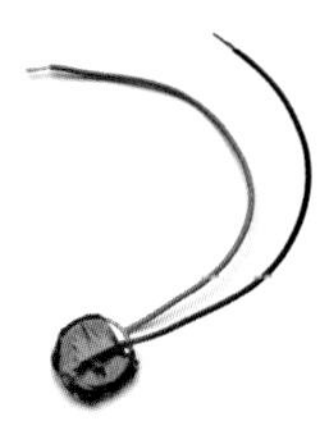

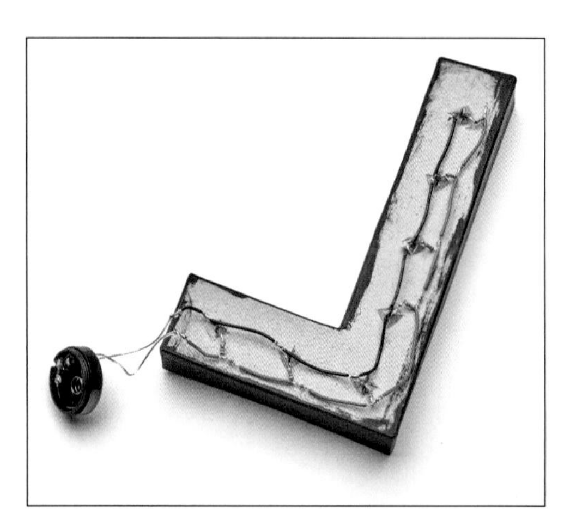

TEST IT! When everything's connected, pop open the battery holder and insert the two batteries. The LEDs inside the letter should glow brightly.

13. When everything's working, use the hot-glue gun to touch a drop of hot glue over each solder connection to protect them.

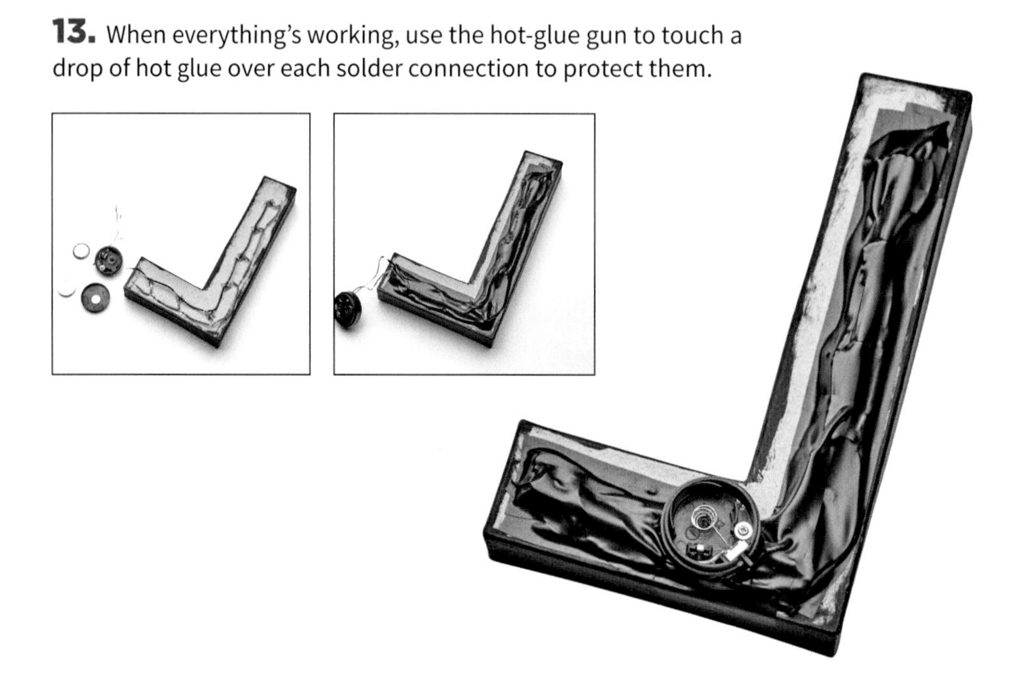

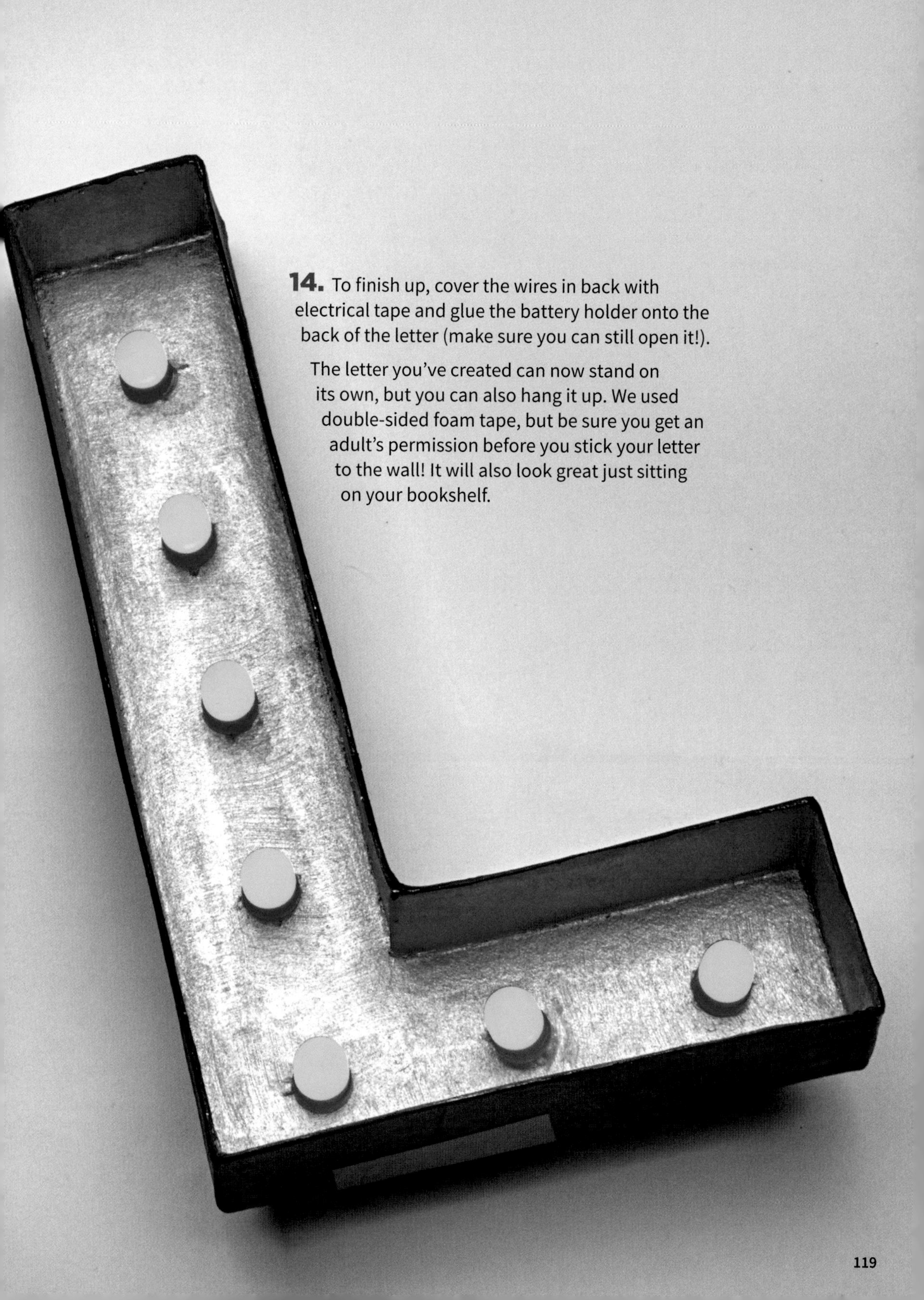

14. To finish up, cover the wires in back with electrical tape and glue the battery holder onto the back of the letter (make sure you can still open it!).

The letter you've created can now stand on its own, but you can also hang it up. We used double-sided foam tape, but be sure you get an adult's permission before you stick your letter to the wall! It will also look great just sitting on your bookshelf.

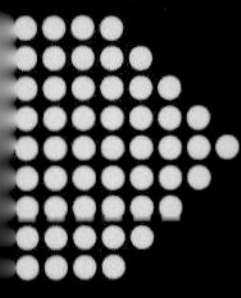

Jazzed-up Jewelry: Start by Making Glue Jewels!

SKILL LEVEL 1

You can't make jazzed-up jewelry without jazzed-up jewels! This quick, three-step process is all you need to make the Glue-Jewel Glimmer Ring, page 122, and the Glue-Jewel Royal Crown, page 125—or invent your own jazzed-up jewelry! Embedding an LED in hot glue diffuses the light, so your Glue Jewel will shine with a soft glow.

Get your tools & materials...

TOOLS

- Black marker
- Hot-glue gun and glue sticks
- Silicone candy mold
- Optional: craft knife or scissors

MATERIALS

- Several LEDs (your choice)
- Toothpicks
- Tape (any kind)
- Optional: glitter, paint, food coloring, and so on

...and MAKE it!

1. With the black marker, darken the long positive (+) leg of each LED. Then wiggle a toothpick crossways between the LED's legs. Tape the legs in place, firmly enough to hold the toothpick, and set aside.

2. Using the hot-glue gun, fill a mold shape in the silicone candy mold with hot glue. (Go slow to avoid trapping air bubbles in the glue.) Let the glue set for a minute, and then gently poke a toothpick-topped LED into the center, legs sticking up. The toothpick will keep the LED upright and in place.

3. When the glue is completely set and cool (about 10 to 15 minutes), gently pop your glue-jewels out of the mold. Remove and discard toothpicks and tape. If any jewel edges look messy, just trim with a craft knife or scissors.

Make as many Glue Jewels as you want! If you want to decorate them before they dry, try swirling in a tiny pinch of glitter or a drop of colorful paint or food coloring before setting in the LEDs. Experiment to see what you like best.

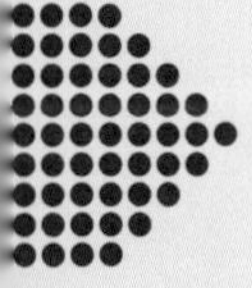

Glue-Jewel Glimmer Ring

SKILL LEVEL 2

Add a touch of brilliance with a glowing Glimmer Ring.

Get your tools & materials...

TOOLS

- Paintbrush
- Drill with 3/16" bit
- Black marker
- Craft knife or scissors
- Hot-glue gun and glue sticks
- Optional: craft knife

MATERIALS

- Acrylic paint (any color)
- Cap from a 1-liter soda bottle
- Elastic string
- One round Glue Jewel (page 121)
- Copper tape
- One 3V coin-cell battery (CR2032)

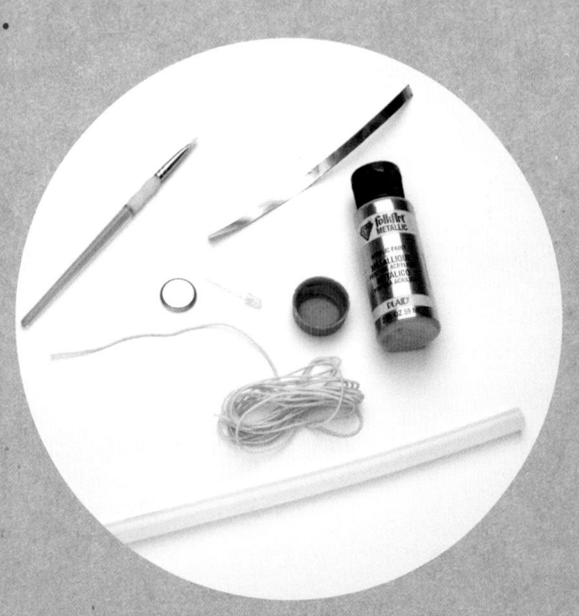

...and MAKE it!

MAKE THE RING BASE

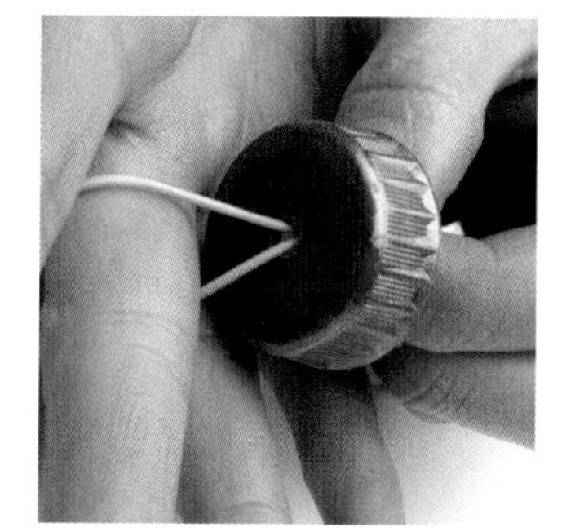

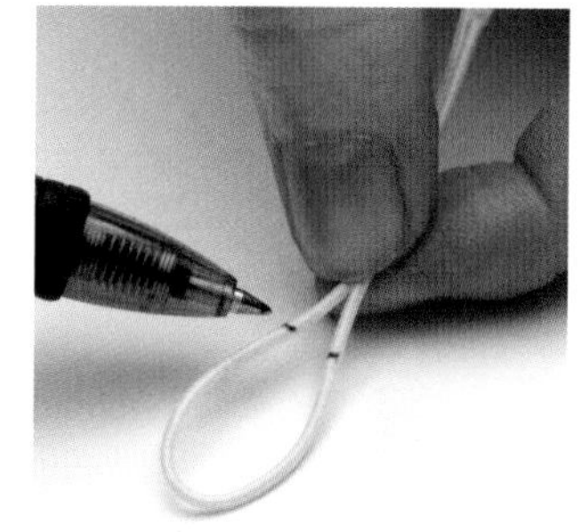

1. Using the paintbrush and acrylic paint, paint the sides of the bottle cap. We chose gold, but you can use any color you want. Then, using a drill with a 3/16" drill bit, drill a small hole in the center of the cap.

2. Feed a loop of elastic through the hole and put it around your finger. Pinch the elastic together inside the cap and use the black marker to mark your ring size.

3. Using scissors, trim the ends of the loop, and using the hot-glue gun, glue the loop ends inside the cap so they sit as flat as possible against the bottom.

ADD YOUR GLUE JEWEL

4. Check to be sure the round Glue Jewel fits tightly into your bottle cap. A Glue Jewel that's too loose could fall out. You should just be able to wiggle it out so you can turn it off by disengaging the battery when you're not using it. If it's too small, add some glue around the edges so it fits in better. If it's too big, use the craft knife or scissors to trim it down to size.

5. When you have a good fit, spread the legs of the LED so they stick out opposite sides of the Glue Jewel. Then, using the black marker, darken the LED's positive (+) leg, so you can easily see it.

6. Using the hot-glue gun, cover the jewel's center with hot glue to keep the legs of the LED from touching. Try to keep the surface as flat as possible. When dry, carefully flatten the darkened positive (+) leg on top of the jewel and fold the negative (–) leg straight up, as shown.

7. Use the copper tape to attach the bumpy, negative (–) side of the battery to the negative (–) leg of the LED. The battery's smooth, positive (+) side should face the back of the jewel, so it will make contact with the positive (+) leg of the LED when the ring is put together.

TEST IT! When you press the battery down against the jewel, the LED should light up. If it doesn't, check to be sure you've got the battery oriented the right way: negative to negative and positive to positive.

8. Snap the glowing jewel into your ring base, fit the ring to your finger—and shine on!

Glue-Jewel Royal Crown

SKILL LEVEL 3

The instructions here are for a queen's crown, which has a peaked top like a tiara. It's easy to adapt these steps to make a king's crown, which is the same, except that all the jewels should be at the same height.

Whether you make a crown for a queen or a king, this royal headgear rules.

Get your tools & materials...

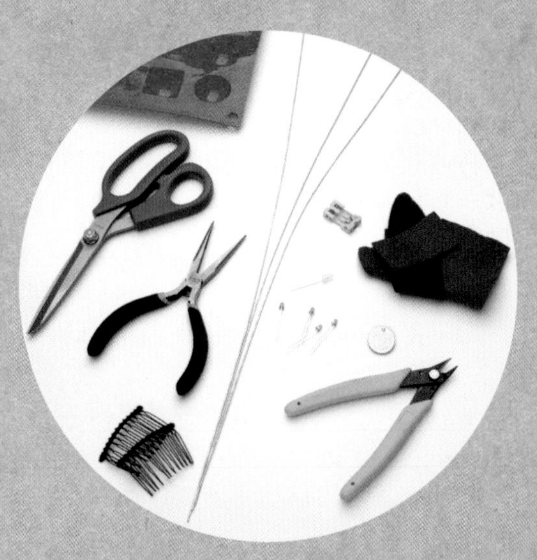

TOOLS

- Black marker
- Ruler
- Wire cutters
- Needle-nose pliers
- Third-hand tool
- Soldering iron and solder
- Hot-glue gun and glue sticks
- Wire strippers
- Scissors

MATERIALS

- Three 18" lengths of 18-gauge or 20-gauge uncoated floral wire, unpainted, silver only (buy pre-cut 18" lengths in craft stores, or purchase longer lengths and cut to size using wire cutters)
- Five Glue Jewels (page 121)
- One 3V coin-cell battery (CR2032)
- Two scraps of felt, each about 2" square
- Hookup wire
- Battery holder for 3V coin-cell battery (CR2032)
- Two 2" plastic hair combs or other hair-pins (like bobby pins or hair clips), as preferred

...and MAKE it!

MAKE THE FRAME

1. Using the black marker and a ruler, mark dots at 5", 7", 9", 11", and 13" on one 18" length of floral wire. This is your "main ground lead"—the negative (–) side of your circuit. Set it aside.

---

2. Using wire cutters, take another 18" length of wire and snip off five pieces: two 1.5" lengths, two 2.5" lengths, and one 4" length. Using needle-nose pliers, bend hooks at one end of each piece. Set aside the extra wire. (To make a king's crown, make all five pieces 2.5" long.)

3. Using the third-hand tool to hold the marked wire as shown, hang the small hooked pieces at each black mark. Place the cut wires in this order: short, medium, long, medium, short.

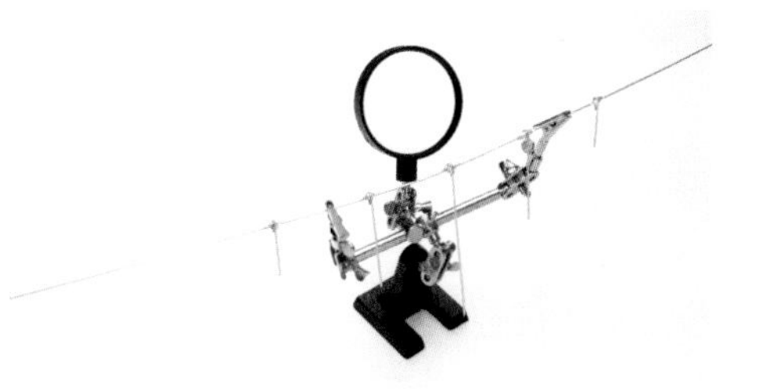

4. Use the needle-nose pliers to squeeze the hooked pieces tight to the long main ground lead. Using the soldering iron, solder each hanging wire in place. Once the solder has cooled, use a hot-glue gun and add a blob of hot glue to each solder point to secure the joint.

WIRE THE GLUE JEWELS TO THE CROWN

5. When the solder and glue on the wire frame are set, lay the frame on a flat surface, solder side up. Set your five Glue Jewels in place, underneath the ends of each clipped wire. Make sure the "crown jewel" (the one you're using at the top) is in the center. On each LED, bend the short, negative (–) leg straight down, so you can attach it to the wire frame. Bend the long, positive (+) legs away, as shown: the two on the right, bend right; the two on the left, bend left; and the crown jewel in the center, straight up.

6. Wrap the negative (–) leg of each jewel around the crown frame. Push any messy ends up behind the jewel so they won't show. Use a soldering iron to solder each joint in place. When the solder is cool, add a drop of hot glue (per step 4) to secure the joints and completely cover each LED's negative lead.

TEST IT! Take your third, 18" length of wire; this is the positive (+) lead. At one end of the crown, sandwich the battery between this positive (+) lead and the negative (–) main ground lead. Touch the positive (+) lead to the positive (+) legs sticking out from each of the Glue Jewels, as shown. Each should light up! If they don't, check the orientation of the battery and make sure the two leads aren't touching.

MAKE THE TOP OF THE CROWN

7. Using the black marker, put a mark at the halfway (9") point on the third 18" length of wire. This is your "main positive lead." It will complete the circuit to light up your jewels.

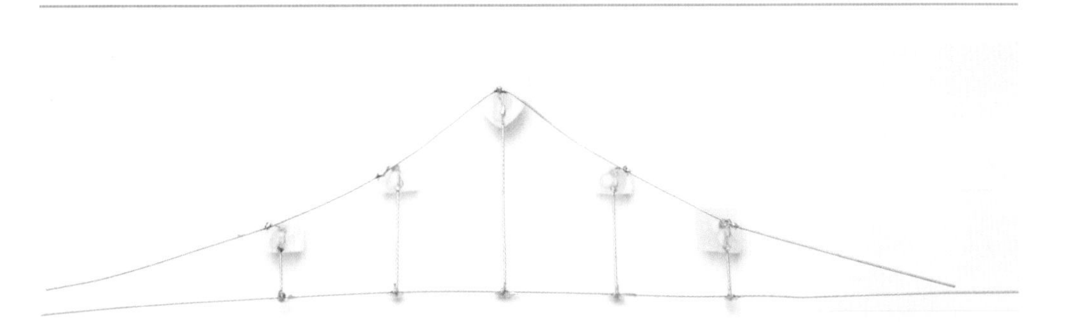

8. At the dot, bend the wire a bit so it follows the line of jewels. Then, at that same center spot, twist in the positive (+) leg of the LED that's sticking up from the crown jewel. Continue doing the same with the other jewels, as shown, wiring in each LED's positive (+) leg to the main positive lead. As before, push any messy knots behind the jewels so they won't show.

TEST AGAIN! Slip your battery between the negative (–) main ground lead and the main positive (+) lead to be sure all your jewels light up. If they don't, confirm the battery's orientation and that the two leads aren't touching.

9. When everything is working, solder each joint; let the solder cool; and then completely cover each knot with hot glue (per step 4).

ADD THE BATTERY HOLDER

10. With the crown in the same position, place two pieces of felt, about 2" square, under the ends of both leads (so the felt will be on the front of the crown). Position the felt so that the bottom, negative (–) main ground lead is about one-third of the way up the felt, which should still be tall enough to reach above the main positive (+) lead. Using the hot-glue gun, glue each felt square to the bottom negative (–) lead only.

11. Choose which end of your crown will have the battery holder (either is fine) and work on that end. Using wire cutters, cut two pieces of hookup wire, each about 2" long. Then, using wire strippers, strip both ends to expose about 1" of the wire inside.

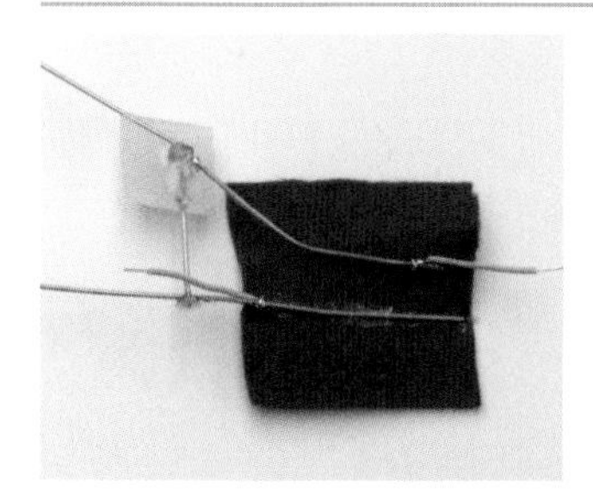

12. By twisting the exposed hookup wire, attach one piece of hookup wire to the positive (+) lead, placing it near the outside edge of the felt. Then attach the other piece of hookup wire to the bottom negative (–) lead, placing it near the inside edge of the felt, and at least 1" away from the wire attached to the end jewel. Using a soldering iron, solder both pieces of hookup wire in place. Once they are cool, bend the two hookup wires away from each other, as shown.

13. At the same end, fold up the bottom of the felt so it covers the bottom negative (–) lead, and tuck any extra felt under the top positive (+) lead. Then fold down the top of the felt over the positive (+) lead. Using a hot-glue gun, glue the felt in place. This will keep the two leads from touching each other and shorting out the circuit.

14. Turn the crown front side up. Place the battery holder on top of the felt (the holder goes on the outside so it doesn't rub against your head). Align the battery holder so that the positive (+) side (marked on the inside) faces the hookup wire attached to the upper positive (+) lead, and the holder's negative (–) side faces the hookup wire attached to the bottom, negative (–) lead. Using the soldering iron, solder the hookup wires to the battery holder, and let cool.

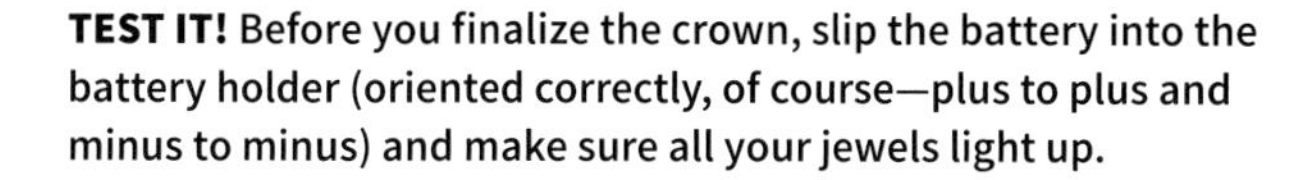

TEST IT! Before you finalize the crown, slip the battery into the battery holder (oriented correctly, of course—plus to plus and minus to minus) and make sure all your jewels light up.

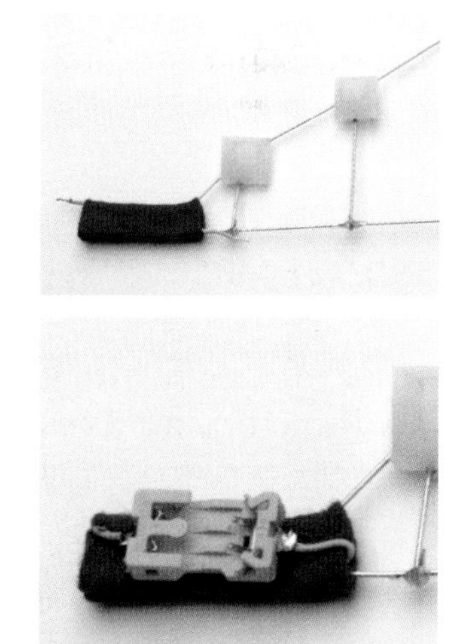

15. When everything's working, use a hot-glue gun to glue the battery holder's edges to the felt, to keep it in place.

GLUE ON THE HAIR COMBS (OR HAIRPINS)

16. Almost done! Turn the crown over and work from the back. On the side without the battery holder, fold up the bottom of the felt to cover the bottom negative (–) lead, and tuck any extra felt under the top, positive (+) lead. Fold down the top of the felt over the positive (+) lead, and using a hot-glue gun, glue the felt in place.

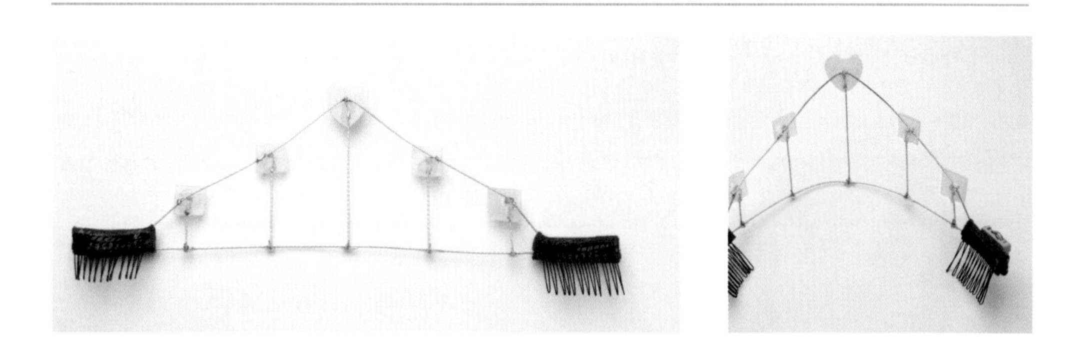

17. Using a hot-glue gun, glue one hair comb (or hairpin) onto the felt on each end. Then, using scissors, trim the felt as necessary to neaten everything up. Bend the whole crown gently to fit your head...and that's it!

Ready? Slip the battery into the battery holder and congratulate your royal self on a job well done!

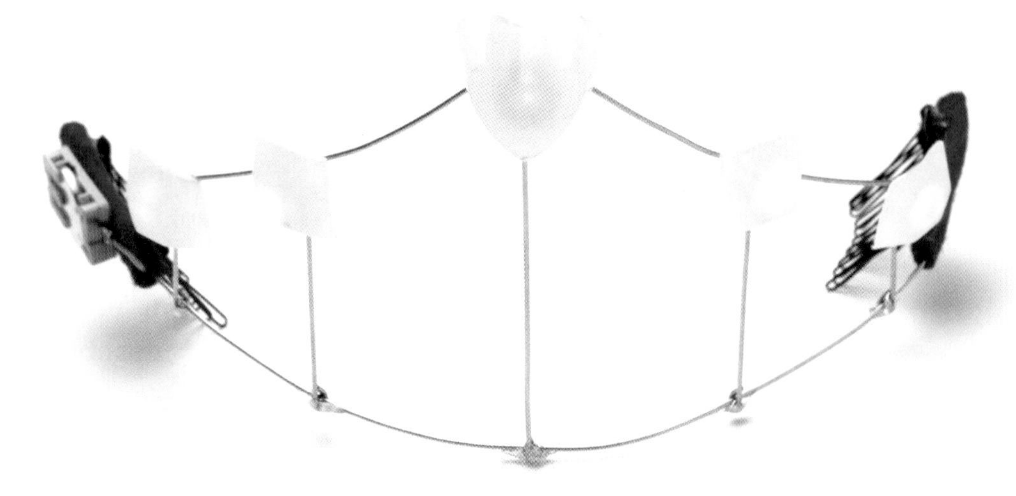

LEDs are amazing, but if you want to expand your repertoire, check out EL wire, a flexible wire that glows in neon colors when electricity passes through it.

The term "EL wire" is short for "electroluminescent wire." Most people pronounce it "E-L wire," but some people say "el wire"; either is fine.

EL wire comes in different forms—from thin wires and thick tapes to whole panels—and in every color you can imagine. It's driven by a controller called an inverter, which changes the power coming from your batteries into the high-voltage alternating current the wire needs to glow. But don't worry—even though it's high voltage, it uses very little current, so it's safe for you to use and wear, and it's great for creating personalized projects.

You can attach EL wire to just about anything, from backpacks to bicycles, but it's especially fun to use on glow-in-the-dark hats, shoes, and one-of-a-kind fashions. You can braid different colors of wire together to create special effects, cover the wire with bits of electrical tape to design patterns, or use special controllers to make the wire blink, pulse, react to sound, or even look like it's moving! Let your imagination be your guide.

MORE TIPS & TROUBLESHOOTING

FOR ALL YOUR LED PROJECTS

- Get in the habit of using a black marker to darken the long, positive (+) legs of all your LEDs, and you'll always be able to track them—very handy when you start making complex circuits.
- If an LED doesn't light up right away, you probably have the legs on the wrong sides of the battery. Flip the battery and try again. If it still won't work, try using a new battery or LED.
- Handle your batteries with care. Don't jumble them together, since this can short them out, and dispose of them responsibly when you're done. Don't just throw them in the trash! See page 9 for more.
- You can collect LEDs—and repurpose them—from lots of unexpected, inexpensive objects. Look for reusable strings of LED holiday lights, broken toys, or key-chain-sized mini flashlights (some have flash settings). Check out your local hardware store with your Maker brain, and watch for new opportunities all around you.

FOR GLOWIE/THROWIE PROJECTS

- If you want to make a Glowie or Throwie last longer, you can make a simple on/off switch from a thin strip of stiff plastic. Just insert the plastic between the battery and one leg of the LED (either leg will work) before you tape it all together. The plastic strip will interrupt the circuit and keep the LED from turning on. Pull the plastic out when you want the LED to come on; push it back in when you're done.

FOR PAPER-CIRCUIT PROJECTS

- If a circuit-sticker LED is not lighting up, check that it's oriented correctly: the wide, positive (+) side on the positive (+) lead, and the pointy, negative (–) side on the negative (–) lead. If that doesn't solve the problem, be sure the LED is stuck tight to the copper tape.
- Remember to bend the copper tape around corners as you're making each circuit. Don't cut copper tape if you can help it.
- If the copper tape gets broken or cut in the wrong place, put another piece over the break. Be sure the pieces are stuck tight to each other to maintain the circuit. Do the same thing if you discover that you need to patch another piece over a too-short length of copper tape.
- Don't let negative and positive lines of copper tape touch or cross one another, or you'll short out the circuit. If one line of copper tape must cross another, cover the bottom line with a scrap-paper bridge to keep them from touching.
- Copper tape will still work even if it gets covered in glue or transparent tape.

+ Don't have any circuit stickers? You can use regular LEDs. Just separate their legs and tape or glue them securely into the copper-tape circuit. Remember that the LED's long leg is its positive (+) side and the short leg is its negative (–) side.

FOR SOFT-CIRCUITRY PROJECTS

+ Before you start a project, be sure your sewing needles will fit through the components. Some components have holes too small for certain needles.

+ Sewing with conductive thread can be tricky. Even when you're working carefully, it can get snarled and knotted. If your thread breaks, or if you need to cut and start over for any reason, you can reconnect by sewing a new piece tightly around the broken end. Do the same thing if you accidentally begin sewing with a strand of thread that's too short.

FOR MIXED-MEDIA PROJECTS

+ In commercial electronics, wire colors are coded for different kinds of uses, but you don't need to worry about that in these activities. Hookup wire is hookup wire. Using different colors just helps you keep track of your circuit. Red normally indicates positive (+) connections; black normally indicates negative (–) connections.

ACTIVITIES BY SKILL LEVEL

SKILL LEVEL 1

SKILL LEVEL 2

SKILL LEVEL 3

ABOUT THE AUTHORS

Emily Coker is a maker professional who is passionate about empowering herself and others through hands-on learning and making. When not creating projects for the masses, she can be spotted tinkering around in her shop futzing with electronics, robotics, and the latest in tech and craft. She is an avid comic and graphic novel enthusiast who enjoys drawing. Emily is also a contributing author for *Make:* magazine and Makezine.com.

Kelli Townley is a lifelong creator and tinkerer. When she was a little girl, she built circuits with her dad, sewed fingerless gloves with a 1930s Singer sewing machine from a thrift store, and disassembled electronics (like that VCR that kept eating tapes). As an adult, she's worked in video games, VFX/animation, education, and most recently, virtual reality. In her spare time, she's constantly doing hands-on projects like upholstery, cosplay, woodworking, and quilting. Kelli is passionate about encouraging makers of all ages and skill levels, and hosts frequent making get-togethers with her friends and colleagues.

ACKNOWLEDGMENTS

Maker Media wishes to thank the many people who contributed to the book, especially the contributing Makers Kelli Townley, Emily Coker, Angie Callau, and Jie Qi. A huge thank you to the book team, including book packager Leslie Jonath, writer and editor Ruth Tepper Brown, photographer Rory Earnshaw, designer Kevin Plottner, managing editor Dean Burrell, and copy editor Jeff Cambell as well as to Evan Brown, Katje Richstatter, Peter Cole, and Eunice Choi and to readers Natalie Freed, Ken Murphy, and Pat Murphy.

HOW TO CONTACT US

Please address comments and questions concerning this book to the publisher:

Maker Media, Inc.
1160 Battery Street East, Suite 125
San Francisco, CA 94111
877-306-6253 (in the United States or Canada)
707-639-1355 (international or local)
bookquestions@oreilly.com.

Maker Media unites, inspires, informs, and entertains a growing community of resourceful people who undertake amazing projects in their backyards, basements, and garages. Maker Media celebrates your right to tweak, hack, and bend any technology to your will. The Maker Media audience continues to be a growing culture and community that believes in bettering ourselves, our environment, our educational system—our entire world. This is much more than an audience, it's a worldwide movement that Maker Media is leading. We call it the Maker Movement.

For more information about Maker Media, visit us online:

- *Make:* and Makezine.com: **makezine.com**
- Maker Faire: **makerfaire.com**
- Maker Shed: **makershed.com**